T0140387

Studies in Systems, Decision and Control

Volume 156

Series editor

Janusz Kacprzyk, Polish Academy of Sciences, Systems Research Institute, Warsaw, Poland

The series "Studies in Systems, Decision and Control" (SSDC) covers both new developments and advances, as well as the state of the art, in the various areas of broadly perceived systems, decision making and control—quickly, up to date and with a high quality. The intent is to cover the theory, applications, and perspectives on the state of the art and future developments relevant to systems, decision making, control, complex processes and related areas, as embedded in the fields of engineering, computer science, physics, economics, social and life sciences, as well as the paradigms and methodologies behind them. The series contains monographs, textbooks, lecture notes and edited volumes in systems, decision making and control spanning the areas of Cyber-Physical Systems, Autonomous Systems, Sensor Networks, Control Systems, Energy Systems, Automotive Systems, Biological Systems, Vehicular Networking and Connected Vehicles, Aerospace Systems, Automation, Manufacturing, Smart Grids, Nonlinear Systems, Power Systems, Robotics, Social Systems, Economic Systems and other. Of particular value to both the contributors and the readership are the short publication timeframe and the world-wide distribution and exposure which enable both a wide and rapid dissemination of research output.

More information about this series at http://www.springer.com/series/13304

Sayed Hadi Sadeghi

Pathology of Learning in Cyber Space

Concepts, Structures and Processes

 Springer

Sayed Hadi Sadeghi
Faculty of Education and Social Work
University of Sydney
Camperdown, NSW
Australia

ISSN 2198-4182   ISSN 2198-4190 (electronic)
Studies in Systems, Decision and Control
ISBN 978-3-030-08258-1   ISBN 978-3-319-91449-7 (eBook)
https://doi.org/10.1007/978-3-319-91449-7

Printed on acid-free paper

This Springer imprint is published by the registered company Springer International Publishing AG part of Springer Nature
The registered company address is: Gewerbestrasse 11, 6330 Cham, Switzerland

To my Dad Martyr Sayed Ali Sadeghi who taught me love and faith,

To my kind Mom who sacrificed her life to my success and my progress

And

To my Dear wife Zeynab for her loving support and my

little Hero Sayed Mir Mohammad

Contents

1   Concepts 1

2   Cyber Space and Real Space 23

3   Training in Cyberspace 39

4   Learning Objectives in Cyberspace 55

5   Pathology Caused by Cyber Education 65

6   Reduce and Deal with Injuries by Training in Cyberspace 93

7   Conclusions 103

References 109

List of Figures

Fig. 1.1 Relationship between real space and cyberspace   8
Fig. 1.2 Extensive training and learning areas in cyberspace   16
Fig. 1.3 Providing educational content in a cyber-environment   16
Fig. 1.4 Distance learning trends   18
Fig. 1.5 Cyberspace learning paradigm waves   20
Fig. 3.1 The conceptual from work for using email
in useful teaching   49
Fig. 3.2 The stages of the formation of the virtual group   50
Fig. 3.3 Transferring goals and thoughts in the form of images   51
Fig. 4.1 Education process   56
Fig. 5.1 Cyberspace and internet damage to the family   75
Fig. 6.1 Feedback in cyberspace   96
Fig. 7.1 Learning pathology and learning in cyber Space   105

Introduction

The world is moving at an accelerating pace. The pace of these changes is such that one moment can quickly hold back communities and governments alike from the rest of the world. On the other hand, the lack of full knowledge and analysis of these developments as well as the formation of new technologies can cause backwardness in unknowing societies, and irreparable damage to the government society, the individuals, and so on. In this regard, as an expert says:

> With this acceleration, you can only keep you where you are. If you want to go elsewhere, you have to double your acceleration at least.

After the revolution that took place in the United Kingdom under the name of the Industrial Revolution, which formed the basis of the formation of the industrial society, in the last quarter of the twentieth century, a much more important technological revolution took place around information and information technology called the Information Technology Revolution, or the Information Revolution in United States. This revolution has transformed the "industrial society" into what it calls the "information society" or "network society." As a result, this has led to the formation of a paradigm with the title of two globalizations among individuals and scholars. This paradigm implies two kinds of worlds in our day-to-day living space: human beings. The first world is "real space." The world that was already in existence, and human beings have never played a role in it, and our past have only experienced this kind of world. The second world is the "cyber space" or "virtual space" that does not last for more than a decade. This world is the result of the revolution in human communication and information technology. This cyberspace, called the second world, places humans in the face of virtual reality. In the virtual world, humans face a boundless, multicultural, yet unified environment. The cyberspace is the result of the transformation and development of the power that combines different aspects of our daily lives and exposes human beings to new forms of interaction that no longer depends on the attendance of the audiences in the same place. In this regard, electronic communications and digital networks are changing the structure of work as well as the forms of personal communication and entertainment of human beings, and have created a space beyond the real space in the Internet environment called "cyberspace." This newly formed space during the process of electronic developments has caused the world to interfere with the Internet through the Internet and be close to each other at far distances. It has also created many opportunities and facilities at various levels of the world, government, local, and even individual. However, it has created limitations and challenges at different levels as well. These changes are due to the formation of a cyberspace to a degree that has transformed all human activities according to especial aspects of a cyberspace. On this basis, it can be said that any human activity in real space has been shaped electronically and shadow in the cyberspace, causing the real space and cyberspace to interconnect, in two parts that are affecting each other directly and indirectly.

Holding virtual courses and learning and educating as a whole in virtual space (cyberspace) is a special experience providing information in any field for every person, specially students who are interested in online education or cyber-learning can use this information in a useful and attractive manner to practice and learn based on Internet. So the state of art for this book can be twofold, one for identifying the reading in resources and second easy understanding for readers about the author's main and basic ideas with well-organized heading. So covering all major issues in online learning and training should be provided for anyone that can bring useful web resource and reading list.

Learning and training in cyberspace can bring personal and especial experiences and also knowledge for interesting goal of sense creation of practices and operations based on Internet in using of online education in a useful manner for anyone and any student who is interested in online learning. Therefore, the state of art for this book can be twofold, one for identifying the reading in resources and second easy understanding for readers about the author's main and basic ideas with well-organized heading. So covering major issues in online learning and training should be provided for anyone that can bring useful web resource and reading list.

The most important human activities that are influenced by cyberspace are learning and teaching. Although learning and teaching have undergone some changes in the historical process and the course of time, these changes have not been so great that they can impress the traditional methods of learning and education. But with the formation of the cyberspace and its development over time, learning and teaching have undergone massive changes. These changes have resulted in a quantitative and qualitative development of the learning and teaching processes that have provided various capacities and potentials in this field in different contexts and discussions ways. But most of them are general and in line with the overall benefits through e-learning and e-teaching that is formally carried out in universities and schools through cyberspace. But, unfortunately, none of them fully explores the learning, education, and cyber-training resources which should be used by individuals, groups, organizations, governments, and others in pursuing to achieve their goals. In addition to this, they did not pay much attention to the challenges and injuries caused by learning and teaching in cyberspace. This has led us to look at these issues and some other issues related to learning and training in cyberspace more widely and comprehensively, and also consider them beyond official formal learning and education. Most importantly, the issues such as injuries and challenges, in different ways and in different levels, result from learning and education in this space. Our goal in this book is to go beyond simple and repetitive issues that have been told about the cyberspace and the challenges it poses and is now underlined. Our main audiences are those who are probably the first to face these issues in principle with this new perspective. Although intended readers of this book are scholars and students from the fields of science education, information technology, sociology and educational technology as well as interested parties and related authorities, but reading this book also can be helpful for people who tend to better understand very new topics, such as cyberspace, learning and training in cyberspace, etc., and their relevant issues. We follow four main goals in this book:

- Understanding the cyberspace as a new emerging space, and familiarity with its capacities and limitations, as well as how to manage this space thoroughly.
- Review and analyze the training process and educational resources in cyberspace.
- Study and analysis of procedure of obtaining educational goals in cyberspace.
- Pathology of learning and learning in cyberspace.

Having regard to the above goals, we would inevitably have to pay attention and consider many other topics that are in some way related to the topics covered in this book, but we refrain from bringing them in this book due to numerous more important topics discussed and also keep the book short and clear enough so that the readers can well keep track of the main themes of it. Based on this, the content of the book is divided and arranged into seven chapters. The book begins with discussions about the main concepts of cyberspace, the relationship between cyberspace and real space, learning and education. The relationship between cyberspace and real space is described, and capacities, judiciary, and concepts related to cyberspace are presented. In the following, cyber curricular education forms are described in terms of teaching and learning resources in cyberspace. The discussion presented in this book consists of two main sections: In the first section, we outline the objectives of training in cyberspace at different levels, and in the second section, we describe the injuries caused by learning and training in the cyberspace at different levels, and then a general attention is drawn to show how cyber-training is handled and receives feedback and, finally, we provide a summary of the topics presented in the book (Fig. 1). The author's main strategy, on the one hand, is the provision of basic aspects and, on the other hand, presentation of the main content with a simple approach. Although it was difficult to achieve both of them, the explanation of the content in the conceptual framework and the alternative approach was less easy to write, but I hope that this work will be for a wide range of enthusiasts, with different intellectual, specialized, and executive backgrounds, including academics, policymakers, managers and planners of educational and cultural units and institutions, and so on. The book might be flawed in terms of its content and the presentation style and the author hope that all readers,

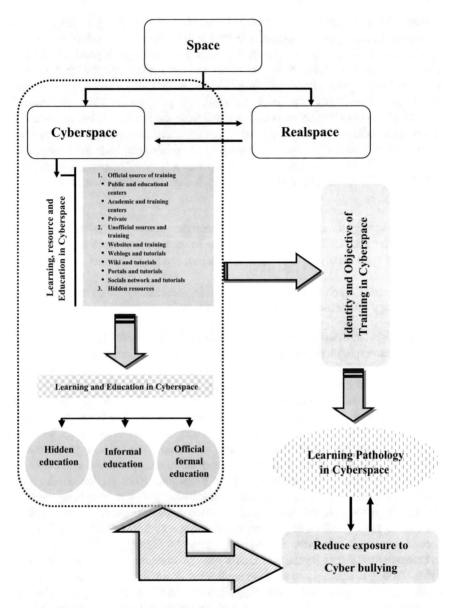

Fig. 1 The whole concept of the book

especially the scholars of the field, with a dim view or with their elaborate guidance, will indebt him with magnanimity. Thanks and gratitude to all those who helped the author in the various stages and provided the conditions for writing and publishing it.

Chapter 1
Concepts

Introduction

In the last decade, several social media such as Facebook have been emerged as the popular social networking, websites in internet through world. There was also noticeable of speed increasing in international expansion of these social media, networks, and websites and in a word cyberspace (Graham, Thompson, Wolcott, Pollack, & Tran, 2015).

So changing in traditional classroom training and learning (traditional education) into online education had been affected the aspects of educational practices (Hong, Tai, Hwang, Kuo, & Chen, 2017). Therefore in online learning, there are instructors can no longer see the learners for dynamic learning, face to face communication and training directly about tasks and duties. In this regard, there will importance to state this environment as the main path of future learning, so by knowing that these days you hear many new emerging words related to technology and cyber space. Before we begin to know how to work with these words, we should know their correct meanings and current conceptions. This is because it is not possible to handle something without having a complete knowledge about it. During this chapter, we will focus on current conceptions of cyber space and information technology, which exist in society, and also the relationship between cyber space and real space. The concepts of training and quality and best practice of technology facilities to help learning system are also stated and discussed.

© Springer International Publishing AG, part of Springer Nature 2019
S. H. Sadeghi, *Pathology of Learning in Cyber Space*, Studies in Systems,
Decision and Control 156, https://doi.org/10.1007/978-3-319-91449-7_1

Cyber Space

Information Technology

Information technology (IT) is a combination of two words of information and technology. First, the conceptual definition of these two words is examined separately.

Information: The term "information" is derived from the "notification". Its Latin root is the word "informer man", which means "shaping" and "molding". This word is used for the first time in Cisro and is synonymous with "teaching" or "knowledge transfer". The word information was invented in the late middle Ages and has come to the fore in the second half of the twentieth century. Researchers looks at the state of structural information. The main structure of information consists of the relationship between the sign, the thing and the individual. This connection is expressed in this way: "The person is informed by something of a sign." The three components of this association have different names. The "individual" identifies the receiver of the information as a listener, reader, or observer, and the "message" symbolizes that "thing" as a content or information message. Researches shows that information dividing to three groups: natural, cultural and technological. Natural information is a report about reality, that is, a person naturally and without mediation is the subject of it. For example, if we hear the piece of music directly and in the place of its execution, it is the same natural information. Cultural information is a recipe for manipulating reality, which means that one can capture more and more of it in the realm of music, for example, the musical notes of a piece of music, the instruction that they can be called cultural information. Technological information is a reality record and in some way rival of reality; that is, this type of information records the reality with the use of technology; it can record the reality at the location and the time of its occurrence. In addition, auto supply of technology information program (TIP) is another developed kind of this type for industries that can be useful in online environments and e-systems technologies. Despite these interpretations, we can say that information is everything we deal with and inform us about events, issues, and affairs. This information may be news in the work or life environment. It may be a rumor in the community, or an e-mail program, which is embedded on the computer screen.

Technology: Technology, as well as information, is presented in many definitions, ranging from "the application of scientific methods to solve problems" to "tools or processes that shape the knowledge or art of a culture." Technology is "any human object or process that can be used to convey messages." Technology can be including the knowledge and innovations as phenomena Internet with general purposes (Korzinov & Savina, 2017). In other words, technology is the science of industries and professions, or the collection of technical and industrial terms, or familiarity with technical principles and technical and technological terms.

Given the various definitions of information and technology that have been presented separately by the experts, it can be seen that the scope of the definition of information technology is vast. So, slate refers to any technology that is controlled and controlled by a microprocessor (computer chip). "McCobell's scientific and technological culture" defines information technology as: "The set of technologies that are particularly relevant to the processing, storage, and communication and information technology linkages include all types of computers and communication systems as well as methods and Reproduction and copying of documentation, written matters, pictures, drawings and films, and mass production methods." McCaucell's Encyclopedia of Technology states: "Information is the improvement of various human and organizational efforts, to solve the problem through the design, development and use of systems based on the use of technology and processes that improve the efficiency and effectiveness of information in the situation." As a result, it can be said that information technology means the knowledge, skill, or technical method in the exploitation of information. Information technology is known as the technology of using computers to gather, store, process, protect, or also transfer information. Nowadays, there is another common term information and communication technology (ICT) to work with computers in a world that is called internet. In other words, information technology is the scientific, technical, engineering and management techniques, and the processing and use of information in social, economic and cultural fields (UNESCO, 2003).

Perhaps what is more important than the meaning of information technology is to understand and clarify the concept of information technology. Because the concept of information technology in addition to technical and skill dimensions will include cultural components, given this fact, the human and community considerations, in addition to technical aspects, will include its cultural and educational aspects, and when these components and functions will enter into the lives of every phenomenon it brings social, cultural, economic, and even educational effects. As a result, this phenomenon will not only act as a hardware but also as an effective software, and the depth of its impact will then increase as the tools and have new communication and information concepts and, as a result, civilization will be based on information. In line with these developments and changes, today, information technology has accelerated in the wake of the whole world, and by connecting real and legal people to the widespread global network, the planet has actually turned into a village (Akbari, Moslehi, Fathi, & Bozorgmehr, 2007). Therefore, natural that all traditional models, which were in some way related to the collection, storage, processing, and retrieval and dissemination of information, would be influenced by the phenomenon of modern information technology in the new state of affairs. This phenomena also affected education and there have been extensive developments in this field area. As such, the evolution of information technology has affected all sectors of society, including education.

Information and Communication Technology

Information and communication technology has emerged from the integration of the three areas of information, computer and communications. The computer hardware division and the supplier of the necessary equipment and devices, data, and information as raw materials within the network and telecommunications are the third part, which has the task settling between the other two parts that combines them with information and communication technology. Revolution of social networking and IT are able to both grow non-academic knowledge and guide social Medias to learn as parallel (Dron & Anderson, 2014; Lumby, Anderson, & Hugman, 2014). This technology is a collection of tools and methods used by computers and communication networks to produce, publish, store, organize, exchange, access, retrieve and disseminate information. It can be said that ICT encompasses a wide range of tools, functions and processes in the fields of production, distribution, organization, exchange, access, and dissemination of information. In other words, information technology includes two aspects of hardware and software, each containing a variety of methods, tools and standards. In other view, ICT (information and communication technology) is defined as a set of technical equipment, methods, device and machines, tools and instruments, science and also the abilities for utilization of information in all aspects of communication (such as production, processing, aggregation, recycling, transfer).

In line with these developments, in recent years, the term "information and communication technology" replaced by "information technology". In the twenty-first century societies, the transformation of change through information and communication is evident in all aspects of individual and social life. However, during the history of mankind, the change has always been a human being. But what has become evident in recent decades is that the information and communication technologies have developed and expanded and obtained a lot of speed, quality and ease so that the coordinates of the present, speed and quality and the volume of change that, unlike the past, is unpredictable. Experts and scholars believe that this change cannot be ignored and easily passed by. They also emphasize that we must be fully prepared to face this change that may turn into revolution. They even argue that the most important opportunity for dealing with these changes is to recognize the full facts of them and form an intelligible and fitting education. The capability and capacity of information and communication technology and its resulting developments have been able to address the shortages and deficiencies of traditional education and bring about a great deal of change in education. Using cyber space, new and effective methods of teaching and learning can be achieved. Information and communication technology has created a new type of education and learning that does not require the physical presence of the teacher in the classroom and learning is possible in non-classroom environments, so that learners share their information with other people. And thus use their own information and that of others (Callan, 2009). In sum, ICT is one of the important scientific achievements of recent decades, which has played a significant role in the quality of affairs such as education. Education experts now believe that the use of information and communication technology can bring many benefits to promoting the mission and objectives of teaching and learning as well as improving its quality.

The Process of Forming a Cyber Space

After the revolution that took place in Britain under the name of the Industrial Revolution, which formed the basis of the formation of an "industrial society", in the last quarter of the twentieth century, a much more important technological revolution was rooted in technologies, communication and information called the Information Technology Revolution or Information Revolution. This revolution occurred when it was home to the United States. This revolution has transformed the "industrial society" into what it calls the "information society" or "network society". These changes have led to the emergence of a new paradigm of global-ization. This paradigm implies two kinds of worlds in our daily lives. The first world is the "real world." The world that was already in existence, and human beings have never played a role in it, and our past have only experienced this kind of world. The second world is the "virtual universe" that does not last for more than a decade. This world is the result of the revolution in information and communi-cation technology built by man. In line with these developments, the ground for the emergence of the Internet was provided as an emerging technology derived from information and communication technology. The idea of the Internet was formed in the early 1960s when the United States military organization and Rand Company decided to use computers to ensure that there could be continuous communications in the event of a nuclear invasion. A system with a central computer will be highly vulnerable to intruder damage. What we now use as the Internet is a system of related computers, in such a way that these computers use a large number of telephone lines to communicate with one another, thereby ensuring that, in the face of the inability of a route to send information there would be another route. This prototype of the Internet was created in 1969 under the name Arpanet (Network of Advanced Research Projects Agency) and implemented by the US Department of Defense come on (Di Angelis, 2004; Ghorbani & Ghorbani, 2014). With the advancement and development of this technology, since the early 1980s, the phe-nomenon of computer-based telecommunication was used, and in fact a link was established between computers and telecommunications, which entered into the information age with the network phenomenon resulting from this link. Today, with its unique features, the Internet has had tremendous effects on all legal and political levels of the international community. The increasing Internet development is such that if only a few decades ago, government agencies and research centers and individuals only allowed the exploitation and activity in cyberspace, now even ordinary people can easily access this product. Like the age of information, the Internet, which involves the connection of several computer networks, can go beyond geographic boundaries and creates an environment beyond the physical world in which users can trade, correspondence, buy, consult, study, and …

In the course of developments and the advances made by information and communication technology, human life has changed in different societies, and an environment that is commonly referred to as the "cyber space". This is the environment created by technology, known as the "virtual world". Almost all activities that can be done in the real world can be done and simulate in cyberspace.

The Concept of Cyber Space

Most people start by trying to find the meaning of their words by finding them in a glossary. This is certainly a reasonable approach, but it has its own limitations. Dictionaries provide information about how words are used (or at least about how they are used). They can also quickly provide sentences, but the use of words is often obscure and narrower than the definitions of dictionaries. The same occurs for the word cyber space. In Persian, they translate the word 'cyber' into virtual, which is not an accurate statement of this word. The permissibility against the truth is equivalent to the word virtual in English. But we know that the cyber environment is real and not false and virtual. In fact, the cyber world, though not in a material and tangible form, is not felt, but as the information derived from it cannot be titled virtual, the cyber world cannot be considered to be virtual. On the other hand, it should be noted that, in computer science cyber means to go beyond the limits of the physical world. A virtual university is a university that does not have physical buildings and instead provides classes through the Internet (Dolly Goldenberg & Carroll Iwasiw, 2005). In sum, here, in order to be able to express the general concept of the word, we present it as a cyber space.

The term "cyber space" was first introduced by William Gibbs Wander in the 1984 novel Nirmonenser. This writer describes this space as the home of data and information in a future dark circle. After that, the word was widely introduced into the world's vocabulary. Gibbs' cyber space was considered to be an environment where electronic activities were carried out within it. It turns out that in this space, what is experienced is real, such as face-to-face talking with someone or library research or purchase. People can marry in virtual space or obtain university degrees or even order their goods to be delivered at their place of residence. In other words, cyberspace can be "a space in which various activities in dimensions of data and communication, communication and service delivery, management and control can be done through electronic and virtual mechanisms". One of the experts in the field of communication, the cyberspace is a widespread global network that connects to each other various computer networks of various sizes, and even personal computers, using various hardware and software, and with communication contracts. Telecommunications technologies form the basis of cyberspace. Although some of these technologies, such as telegraph and telephone, were invented in the early nineteenth century, but the pervasive and cheap technology of these technologies, which is the main prerequisite for the advent of cyber space, has taken place in recent years. Therefore, computer cyber space is referred to an electronic location where a group of people meet and discuss each other. Technically, the virtual space also means the information space, which is connected with the computer systems of digital networks, which ultimately communicates with the "mother" of all networks, i.e., the Internet. In other words, any non-physical background created by on-line computer systems can be considered as a cyber space.

The Relationship Between Real Space and Cyber Space

With the development of information and communication technologies and theirs changes over time, including the formation of cyberspace, human life has undergone many changes in various ways. The development of this space, which has many capacities and abilities in different grounds, has provided an extensive presence of activities and presence of human beings in the cyberspace. It has also created a virtual reality in this space. The virtual reality, which is a subjective combination of reality and virtual, is based on three fundamental dimensions:

1. Maintenance of the circuit;
2. The possession of the playgrounds of the navigator itself;
3. The prosecution paths are clear that the person's interaction in the cyberspace is as it is normal in the process. The virtual reality is the creation of reality with all its possible capabilities in the "cyber space". This look, creating the real capacities of the virtual world, has evolved into an evolutionary path. All personal and social functions, active and passive actions, presence in loneliness and aggregation, as well as the creation and processing of action in this space are possible and not far away, the period during which the "virtual feeling" is brought to a level of reality and objectivity that is transmitted. It also has odor, taste, energy and "minus body sensation." The capabilities of cyber space are influenced by new information and communication technologies that bring the community into a space that translates it into an "information society". In the new environment, e-government, virtual city, e-learning, etc., which is an expression of their forms in real space, is developing more day by day. In this regard, cyberspace has a variety of effects on communities in real space, which will be in the form of benefits and disadvantages in society. So it can be admitted that these technologies are useful tools and not just some mere concepts, and also can remarkably affect the human communication methods and processes of thought and innovation. Accordingly, in the context of the relationship between real space and cyberspace, the cyber space is by no means a neutral environment. This space consists of several opportunities and disadvantages faced by individuals, groups, society, and governments. In this regard, by gaining an accurate and realistic knowledge of this space, it is necessary to increase the productivity of the opportunities created by this space in various

ways and in different aspects, and to the extent possible, the threats and injuries caused by the cyber space have reduced the country, state, government and society. And in this way, it can prepare itself for future rivalries and advancements, most of which come from the cyber space.

In this regard, it should be noted that virtual space is not something that is separate from real space. Virtual space is a space that relies on the real space and they influence and interact with each other (Fig. 1.1).

Overview

In response to the question "What is the most important and fundamental quality of the universe?" What immediately comes to our minds and has the greatest impact on us is the natural (real) world of the real world: trees, oceans, plants, animals and humans, events and natural phenomena such as earthquakes and warfare and the like.

But this answer may have been valid in the last centuries, but after the British Industrial Revolution, the birth of industrial machinery and tools, and in the years after the information revolution in the United States, which resulted in the birth of a phenomenon called the Internet, a man was faced with a new concept "The virtual world," or "cyberspace", which, despite the fact that it is made by man, has such a tremendous effect on the various aspects of his life that is gradually conquering all the angles of human life. This space, which is based on "information," is building a culture through which this information is in the midst of being empty and changing

Fig. 1.1 Relationship between real space and cyberspace

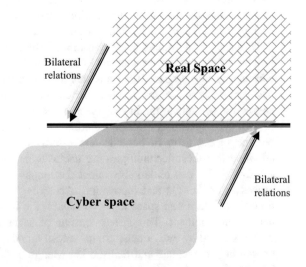

nature. And this new atmosphere, due to these fundamental changes, can affect all our concepts and perceptions of the world around us, and as a result our relationship with ourselves and around us undergoes dramatic changes.

Training

The Concept of Training

In most cases, learning and education are usually considered meaningful among people, while they are not meaningful. Learning is the activity that transmits the foundation of something and in fact supports the learning that is the internal process of converting information into knowledge (Rosenberg, 2005). Unfortunately, in the translation of some of the words and their equivalents in English, there are some mistakes that the word training does not exclude. The turmoil in understanding this term and its close vocabulary is such that many experts, teachers and students of this field have concepts such as education, teaching and they consider one as a worker and use one another. These concepts, although they may have in some ways a common and interlocking aspect, but the concepts are independent and have their own meanings.

Teaching is a premeditated, targeted or targeted action, aimed at providing opportunities to facilitate and speed learning of a learner. There is various kinds of teaching, following tools and methods are among them: teaching, communication, television, internet, etc. the internet learning lead the access, production and processing of information that is available through text, sound, movies, data, documents, etc. (Rolando, Salvador, & Luz, 2013). In other words, the difference between learning and learning is as follows:

Training is activities designed to facilitate learner learning by the mentor or teacher and interacting between the teacher and one or more learners. Of course, we cannot ignore that this definition is just about physical classes. And if we want to give a more comprehensive definition of training, as well as including classroom instruction, the other types of education also include, we can define it as "any predetermined activity or plan which aims to facilitate learning in learners". Therefore, the purpose of education is learning, but a particular type of learning is defined by a particular process and goal (Anderson, Garrison, & Archer, 2004). Now, with the above discussion, it can be said that training is a process in which the establishment of a defined communication between learner and learner leads to enhancement, increase of information, gaining skills and, in general, changes in the stable ratio of abilities mental and practical learning. In fact, we should consider education as a kind of communication, just its goal is to learn, to not only transmit and exchange messages. The role of education is to "design, facilitate and direct the cognitive and emotional processes in order to achieve educational goals that are personally targeted and educationally valuable" (Anderson et al., 2004).

In today countries with labor governments that tend toward education improvement, new opportunities will be emerged for young people and will be made available for them in order of making the mentioned path true (Vali, 2013). As a result of our society needing, the opportunities those are made and offered by new technologies will be useful in education, so these new technologies and learning environment will allow the students to learn on their own way, easy access to information, communicate and accurate evaluation, better solving of difficult problems and etc., so the education will be made more interesting and this education will develop and grow the country.

Training Attributes

Training is known as pre-designed activities to facilitate learning and improve individual performance. The term "training" is used when the learning is to be shaped in a specific way (intentional learning) and the learner is to acquire new knowledge using new knowledge in a particular way or at a certain level of expertise and perhaps within a specified time frame. The training consists of four main elements:

1. The purpose is to improve the efficiency in a particular way. This purpose or purpose is necessarily reflected by the assessment of needs and reflected in the objectives of education and learning.
2. A plan that best integrates educational strategy with learning needs and learner characteristics, and is equally consistent with the measurement strategy that tests the impact of education.
3. Tools and media that are being transmitted by them. It may be a classroom tool, a variety of technologies, a connected study, or a combination of techniques.
4. Evaluation or the ability to present a document on the required success.

Types of training include three categories:

- Formal education

Official or school education includes the part of the educational system that starts from the childhood and continues to the university. This type of organized education is characterized by a distinct structure and follows a systematic hierarchy (UNESCO Advisers, 1990). In other words, it is a process in which the two learning and teaching elements are fully informed and free of charge in the training process. One result is interacting in different formats as the formal education (Vali, 2013). Various formats of interaction are one of characteristics defined as formal education. The goals of formal education, which combines formal education within the framework of advanced education and advanced training within the framework of higher education is provided as follows:

Providing thinking and action tools for the child and adolescence, according to his age, so that he can be flourished and live as a human being and a citizen.

Transferring cultural heritage and providing the necessary cultural resources.

The training of free, informed and responsible humans consider themselves and others to be respected.

Training of first-degree specialists.

Dissemination of Cultural and Educational Culture and Cultural Values of the Society (UNESCO Advisory Group, 2003).

- Informal education

Informal education refers to any type of educational and training activity that is organized, but is realized outside the formal education system. Such as short-term training courses and apprenticeships, in various fields, ideological, political, literacy, health, co-operation, primary care, family planning.

- Tacit continuous training (hidden)

This kind of education is concerned with everything that human beings in their lifetime are structurally unorganized in the fields of knowledge, ability, thoughts, views, and so on. Continuous education through everyday personal experiences takes place with respect to the environment in which the individual grows and works and lives; environments such as home and family, work environment, sports ground, during travel, during study or the use of mass media (such as radio, television, movies), or when they are in cyberspace (which, as a new environment, with high educational capacity, hides behind users) (UNESCO, 1990). Of course, it should be noted that in some cases the sender of the message is unknowingly and in some cases consciously transmits the message and educates the recipient of the message.

In sum, it can be said that the mission of teaching from different angles is debatable. Education generally improves the perceptual and non-perceptual knowledge and skills of individuals. Understanding the realities and processes, the balance of understanding and concepts, and its application, the ability to solve a problem, and gaining the power of analysis and reasoning are among the most important perceptual skills that individuals achieve by learning to achieve them. In addition to acquiring the skills and abilities above, the training seeks to enhance the ability of individuals to learn. An effective learning system is also a way to teach learners how to learn, so that they learn how to learn. This type of learning is very important for the later stages of life of the individuals. The importance of learning a variety of skills is due to the complexities of socioeconomic life in today's world. Some dimensions of human life in the industrial and inter-industrial period are so complex that individuals are required to acquire the skills and expertise required to attend such environments and conditions. Of course, a person is created in such a way that some of the above skills can be obtained through experimentation. But education has two main characteristics in comparison with experimentation: firstly, in the process of teaching, the successful experiences of others, along with the skills and expertise of a scientific analysis, are systematically learned to people. It should be noted that the scope and effectiveness of these skills and expertise compared to personal limited experience are much wider and more comprehensive. Secondly, in the process of experimentation, individuals must be seriously involved with real issues in order to provide them with the opportunity to acquire empirical skills. It is also clear that gaining a skill in this way will be a process of time, but through training, individuals before entering the complex economic and social life, they have learned the skills and are now preparing to enter the environment. Undoubtedly, using the method of education, with its main focus on adolescence, adolescence and youth (that is, a period when the socio-economic life of the human being has not started seriously), is much more efficient and useful in terms of time.

Learning

Different definitions are given for learning, each of which emphasizes specific aspects of the learning process and ignores other aspects of it. Ellson like many others said: Learning is one of the most important areas in today's psychology and at the same time one of the most difficult concepts to define. However, due to the importance of the concept of learning, the experts have derived various definitions. The most famous definition of learning is as follows:

Learning is called the process of making a relatively stable change in the behavior or behavioral ability that is the result of the experience, and it cannot be attributed to temporary conditions of the body such as those resulting from illness, fatigue, or drug use. The meaning of experience in this definition is the effect of external and internal stimuli on the learner. Therefore, reading a book, listening to a lecture, falling a child, thinking about a subject, and so on, is an experience and can lead to learning. Given what has been said, the topic of intentional learning and casual learning it also turns on. Since the learner's encounter with any kind of experience may lead to learning, there is no need for learning to always have a deliberate aspect. In fact, most of our learning happens accidentally, so-called unknowingly. The above definition can be summarized as follows:

Learning is a relatively constant change in mind or behavior that occurs during experience.

Learning has an important in network era; beside the point that students should be focused on education and their learning environment should provide those learning providers, instructors and teachers are the main role in learning process.

Learning and training or education in online environment by electronic instruments, is the new emergence of learning by involvement of computer and electronic device (such as smart phone) for training, educating.

Ganiya also recognizes learning as a change in the condition or ability of a person that is stable over time and cannot simply be attributed to the process of growth. Teaching and learning are two major activities that human beings face from the beginning of their lives. This, on the other hand, refers to the behavior of the elderly to provide the education of children and adolescents, and on the other hand, indicates the inner need of a person to know and his need to gain values and learn the attitudes and skills that are accepted for life with others. As a member of a set and to meet their physical and mental needs, it is inevitable. The role and importance of education and learning is now evident especially in advanced societies, as the amount of global attention and actions in different countries imply its necessity. Among other things, there has been an increasing amount of education in the countries as an official body. Practical attention has been given to teaching even in elementary cultures. In elementary societies, it is also believed that mankind's way of life, skills of coping with problems, methods of food supply through hunting and fishing, as well as values, attitudes and cultural heritage can be transmitted through education to children and adolescents. In the first place, in this way, every human being, independently deals with his own administration of life, and subsequently becomes an additional force in the service of war and solving collective needs. Therefore, the role and necessity of education in primary and advanced societies does not differ significantly with regard to the need for education and understanding of its role and significance, but it is more than the amount of consciousness associated with planning in improving the quality and development of education. And learning as well as the amount of scientific effort to understand the principles and techniques of effective teaching and learning and its application in solving individual and social problems.

Training and E-Learning

The term e-learning was first introduced by Cross and refers to a variety of trainings that were used for learning through computer, Internet, and local network technologies. This type of training, which was done more often through the computer, is not a new topic. This training will last 30 years. Two major developments have taken place in this area. The first, simultaneously with the introduction of multimedia and second-generation computers, was at the time when the use of the Internet for education was introduced. Mainly e-learning is any type of learning, education, or education that is provided by known computer technologies, especially Internet-based technologies. Expansion of information technology makes the opportunity of handling all of educational affairs over the world. This kind of training could be available by all of media such as CD-ROM, Audio and video training tapes, virtual TVs, internet, and extranets. This training is dynamic and its content is completely up-to-date, and experts in each science act in real time, in which the best and most up-to-date resources are available to individuals. This cyber training is online and you can use it whenever you want and meet your needs.

Various definitions for e-learning have been provided, some of which are:

- E-learning is a term that includes Web-based education and technology education on micro technology.

- E-learning is the provision and management of learning and support opportunities through computer networks and web-based technology that helps people develop and work.
- E-learning is learning, teaching and transferring learning experiences that are especially facilitated by computer-based technologies in general and the Internet.
- E-learning is in fact a kind of distance education in which the learner learns in cyberspace, and there is no physical space and face-to-face relationship between people.
- In other words, the content of education is provided by the learner via the Internet (this means that the content is completely electronic) and the relationship between the educator and the learner/learner is done by the.
- Distance education provided the base for e-learning's development. E-learning can be "on demand". It overcomes timing, attendance and travel difficulties.
- E-learning in a traditional setting may include educational films and PowerPoint presentations (Stockley, 2003).
- A web-based learning ecosystem integrating several stakeholders with technology and processes (Alraimi, Zo, & Ciganek, 2015; Chauhan, 2014).

Generally speaking, "e-learning and learning" refers to a set of educational activities that are performed using electronic tools such as audio, video, computer, network, virtual, etc. In other words, all programs that lead to learning through computer networks, especially the Internet, are called e-learning tutorials. E-learning is the usage of IT for evaluation of information and training or educating, so e-learning is the emergence of a new and modern education paradigm. E-learning includes the web usage for reaching the data, information, knowledge without limitation of time and space (Aparicio, Bacao, & Oliveira, 2014). The range of usages and functions of e-learning word is wide and various including applications such as web based education, computer based training, virtual train or virtual classes, and e-business associations. Phenomena such as internet, intranet, extranet, satellite broadcast, video or audio tapes, dial up televisions and compact discs are the tools included in e-learning tools. In other words, e-learning and e-learning include a wide range of functions, including distance learning by radio and television, computer-aided learning, web-based education, virtual classes and universities, etc. Electronic learning has been rapidly growing with the use of computers and the widespread expansion of the Internet and the provision of infrastructure for the transfer of information, audio, image and the development of electronic libraries and journals. In this type of education, because of the ease of access to resources and educational facilities and the availability of a learning environment all day and night from anywhere in the world, the educational goal is achievable for everyone, everywhere and throughout their entire lifetime. But in this type of education, due to the separation and lack of communication between the other issues, student dropout is 10–20% more than traditional face-to-face education. Various researches that examine the causes of deterioration and how to prevent it and increase the efficiency of e-learning have referred to motivation as one of the most important factors for increasing efficiency (Connolly, Stansfield, & McLellan, 2006). In a large division, one can express e-learning at two levels:

A. Online Learning (Network Based Learning):
    All online programs, activities and services that are provided online (always available) or on-line without interruption via online (online) are online education. In other words, online education is referred to as distance learning, which is provided by a wide range of applied software and technology-based teaching methods, including computer training, the Web, the Internet, as well as school, classroom and Virtual University.
B. E-Learning offline (Training CD and e-book):
    An online tutorial is part of an e-learning program that does not use the Web and generates CDs. Offline learning also has some characteristics, it is an automated or self-taught learning which doesn't need continuous communication with teacher, it doesn't include the limitation of access to educational resources but in the other hand it doesn't involve opportunities like chat and teleconference.

In offline learning by the absence of network connection in virtual environments, learners will gain more difficult access to their courses and contents. So this learning can have the access the information in lack of internet connection.

Learning and Teaching in Cyberspace

Today, education and learning are not limited to classrooms (Zhang & Nunamaker, 2003). Various developments and advances in the field of information and communication technologies, resulting in the formation of cyber space has been able to change the process and how to teach and learn in different ways. So, the first generation of computer-based e-learning was designed to be tailor-made and personal. There was no practical way to integrate other learners or a teacher into an educational experience. But, the technology below making the Internet has changed a lot in this regard. Conversations on messengers, online conferences and e-mail have created a variety of ways to blend technology and education. There are not many definitions of cyber space education and learning, and most often this type of learning and learning is synonymous with e-learning and learning. But it should be noted that these two are different and their ratio is total to component. In the context of the difference between "e-learning and learning" and "learning and learning in cyberspace," it should be noted that cyber education and learning is a subset of e-learning and education (Fig. 1.2).

This type of training is for education and learning through the Internet. In this type of training, educational materials are provided by the Web server and via the Internet. This form of education, along with the benefits of computer-based education, also provides higher and wider educational benefits. This kind of training is

Fig. 1.2 Extensive training
and learning areas in
cyberspace

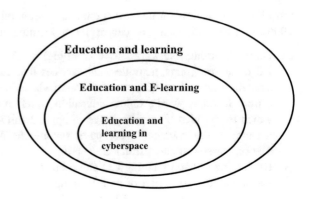

synonymous with Online Learning and Learning. The main ability of education and learning in the cyberspace is beyond its accessibility to information and is based on interactive communication capabilities of both parties. This makes the learning and learning environment in a cyber space an environmentally dynamic and active. And create high-level attractions and capacities in the area of recruitment and education for individuals and users in cyberspace.

In sum, cyber training and learning can be provided as an instruction by the computer at a distance by the teacher and at other levels informally by organizations, groups, governments, individuals, Hidden through games, pictures and … both at the national level and in the wider world. These types of learning and teaching systems have features such as audio, image, text, chat rooms, e-mail, online discussions, assignments, quizzes, etc. (Fig. 1.3). Because of these benefits, many institutions and educational organizations, private sectors, governments, parties and political groups, etc. are investing in implementing these systems for

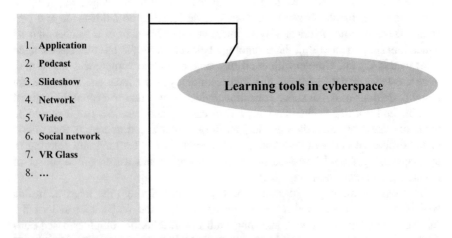

Fig. 1.3 Providing educational content in a cyber-environment

teaching and expressing thoughts and beliefs as well as access. Their educational goals are based on cyber-space capabilities.

Many education scientists have been talking about "changing education paradigm" (Robinson, 2010) recently, pointing out that the power of technology has caused fundamental changes in all aspects of our lives, including education process. To reflect the upcoming changes is important as the need to modify the standardized type of learning and teaching has been growing (El-Hmoudova, 2015)

In a general conclusion, education and learning in cyberspace can be any type of activity that has a specific purpose for influencing users to achieve the desired goal, either through formal centers and organizations such as education and universities, the private groups and informal groups.

Overview

In learning psychology, it is emphasized that learning is the result of experiential learning. The concept of experience in the definition of learning is the interaction between external and internal stimuli and learner. In other words, any kind of interactions between the learner and what constitutes his environment at learning time will include learner experience.

In the farther past, learning experiences consisted mainly of limited activities that people in their surroundings carried out directly and, in accordance with those experiences, reached a level of learning. With the invention of the printing industry and the advent of science and the establishment of institutions such as school and university, knowledge, education and learning were new concepts and became more diverse and more complex.

With the advent of the information revolution and the invention of computers, human beings are faced with a new environment that does not have many features of the traditional past of the past. As a consequence of the religious formation of the new atmosphere, new experiences of learning for humans have different characteristics than the past.

The birth of terms such as "e-learning and e-teaching" and "learning and teaching in cyberspace" are the same type of new experiences and human learning in the present century. This kind of learning and teaching, in comparison with past capabilities, has a greater variety, attractiveness and dynamism for the transfer of concepts and meanings to learners.

E-Learning Generations

From the first, e-learning was defined as distance education so, e-learning and distance education have been connected in the literature historically. It can be said that distance education history dates back about two centuries. In that time there

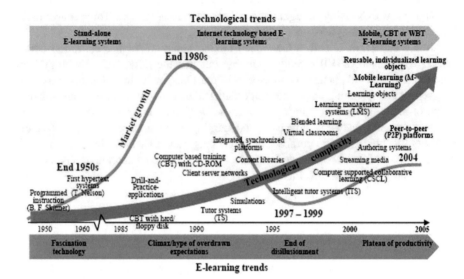

Fig. 1.4 Distance learning trends

was a need to develop correspondence system education to get a geographically dispersed population so, distance education was created. Distance education includes various types of delivery media, learning and training technologies and online instructions. In some ways it can said that creation of distance education and appearance of new technological methods and tools are connected together. Figure 1.4 shows various historical developments that can be conjunct to development of distance education (Sadeghi, 2018).

First Generation: The Correspondence Model

The era of correspondence model is considered as the first generation of distance education, using printed media was common in this era and had a significant effect on the creation of pedagogical structure of instruction. This period is known as correspondence model. One of the important features of this generation is linear delivery of science, knowledge and information over geographical boundaries where teaching and learning is organized asynchronously and is disparate.

Second Generation: The Multi-media Model

Radio and television were appeared in this era and they were considered as new-found technologies in the domain of distance education. Kanas State College,

Purdue University, and Iowa University were the leaders of TV programming and Production of distance education in 1950 s. In addition to printed media, the use of radio and television as the conductors and providers of training courses was common at the beginning of this era. Different learning and teaching methods, skills and techniques were required by this kind of media so there was a need to change the pedagogical structure of courses.

Third Generation: The Tele-learning Model

In 1980s, the use of television and videoconferencing has gradually expanded; almost everyone had access to TV. It was the time for postsecondary schools to begin presenting videoconference-training courses in different campuses. It can be said that this type of media developed a new pedagogical model to handle the new central medium so called videoconferencing (Sadeghi, 2018).

Fourth Generation: The Flexible Learning Model

The important model created in this era includes more complete capacities of world wide web technologies, this capacities provide a three-dimensional e-learning model means it could be available anytime, anywhere at every places. In this era, students could use internet in their online learning, they could experience an attractive learning and educating course that was nonlinear, interactive, dynamic, and collaborative. Thanks to the internet, nowadays students can learn anywhere, any time and at any place promoting social interaction such as sharing information, collaborative education and Computer Supported Collaborative Learning (CSCL), and situated learning.

Fifth Generation: The Intelligent Flexible Learning Model

The Fifth generation is based on the learning and cognitive science with intelligent flexible strategy. "The fifth generation of distance education is essentially a derivation of the fourth generation, which aims to capitalize on the features of the Internet and the Web". This generation has the potential to provide evidence with a high customized experience compared to the prior generations of online program. Currently this generation is identified as involving different blended strategies. Finally based on the all generation of the E-learning we have waves on process of cyberspace learning. As you can see on below figure the three main waves include traditional offline media, web-based learning and LMS systems and at the end embed cognitive and learning science in virtual reality (Fig. 1.5).

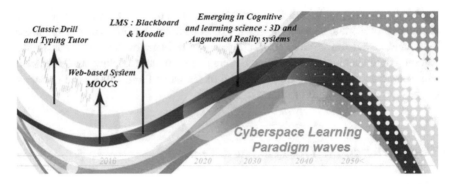

Fig. 1.5 Cyberspace learning paradigm waves

E-Learning Definition

In the recent decades, e-learning was growing so fast causing various definition that consider and address different features of e-learning term and concept, but the feature that are considered by these definitions are mostly the same. We can categorize these definitions, some of them are basic such as E-learning involves using online tools, internet and other technologies to produce education materials, teach students/ learners and organize online learning courses in schools, universities or other teaching and learning organizations. According to some people, e-learning concept deals with various forms of using digital technology methods in pedagogical processes. Other scholars consider e learning as every educational method offering the use of ICT for asynchronous, decentralized content presentation and distribution, as well as for interpersonal communication and interaction (Sadeghi, 2018).

Some authors believe that e-learning refers to the learning and teaching key and address their important frameworks, ICT tools can provide these frameworks in some way to transmit an extensive collection of solutions with the purpose of increasing knowledge and performance. Some reviewers believe that using PC to deliver learning and teaching material forms an important part of e-learning. The focus of other definitions is using tools, applications and equipment included in the processes of learning and teaching where the learner is challenged in a dynamic learning environment. According to mentioned definitions,—"electronic learning" or e-learning provides a wide range of online technologies and methods to create materials for increasing knowledge and performance of education and learning process.

Some of these definitions address social networking communication whereas others focus on electronic tools and applications that are used in the in the processes of learning and teaching where the learner could be challenged intellectually in a dynamic environment. Look at following example:

E-learning is presented by The Open and Distance Learning Quality Council in the UK as 'an effective learning process developed by digitally joining delivered content with support and services'.

'E-learning is an information and communication technology-based learning, pedagogical interaction between students and the content, students and the instructors or among students through the web are the processes appeared in this type of learning'.

E-learning at the Ministry of Communication and Technology of New Zealand is presented as "… the kind of learning makes education easy to get by the use of digital tools and content", it has some kind of interactivity and -online interaction between the learner and their teacher or students may include in it.

Online Learning Consortium in 2015 provided e-learning definition based on two central characteristics; "both of the course level and the program level are included in definitions"; "definitions are consist of three key parameters: instructional delivery mode, time, and flexibility" (Mayadas, Miller, & Sener, 2015). E learning in these characteristics was defines as a new model based on pedagogical oriented definitions. So in this regard, e-learning addresses to "pedagogical processes using ICT to deliver both synchronous and asynchronous learning and teaching affairs". Accordingly E-learning is defined as ICT, computers and networks used to support learners to enhance their learning and educational processes.

Tools used in the new environment appeared in the digital era are the central focus of many e-learning definitions, these tools develop student-centered learning and educational practice, providing new more flexible learning methods (Shopova, 2011). Because some definitions are more focused on the technology rather than pedagogy, scholars said that "if we describe e-learning in terms of the enabling

Table 1.1 Pedagogical dimensions of e-learning

Pedagogical dimension	Type I e-learning delivery-centered paradigm	Type II e-learning learner-centered paradigm
Learning mechanism	Showing and telling	Learning-by-doing
Role of technology	Delivery of content and evaluations	Cognitive tools, scaffolding learning, providing feedback, non-linear access to information sources, and supporting collaboration
Type of content	Didactic written texts and multimedia lectures covering factual information	Realistic texts and multimedia cases, 2D and 3D simulations and virtual worlds, dynamic computer models
Role of student	Passively view or receive content	Actively engaged in problem solving, projects, and collaborative activities
Control	Technology	Learner
Learning outcomes	Achievement on objective memory tests of factual information retention	Ability to solve new problems, performance assessments

Jacobson et al. (2005)

technologies this wouldn't be a useful definition because in this kind of definition there isn't any distinction between the types of design features for various e-learning approaches, and more important, the mentioned distinction doesn't place between different paradigms for teaching and learning" (Jacobson, Kim, Lee, Kim, & Kwon, 2005, p. 79). As a result, definitions of e-learning must cover an extensive range of pedagogical dimensions.

Pedagogical dimensions based on types of e-learning are: type I e-Learning is a delivery-centered paradigm (more traditional approaches) and type II e-Learning is a learner-centered paradigm (active engagement in doing activities). Table 1.1 shows the pedagogical dimensions of Type I and Type II e-learning.

As you can see in the table, Type I and Type II e-learning are different in terms of some dimensions. Although learning basic factual information is addressed in the pedagogical dimensions of type I, but the central focus of pedagogical dimensions of type II is developing deep understandings of the content and problem solving involving learning-by-doing (Jacobson & Spiro, 1994). You can see that Type II e-learning systems provide advanced research practices means exploring ways to design and use advanced technologies to support learner-centered pedagogies.

Chapter 2
Cyber Space and Real Space

Introduction

With regard to social lifestyles of human beings, we should always have some inhibiting frameworks in order to prevent occurrence of damage to each other. Cyber space has become a virtual place in which real human beings are now interacting just as they interact in real spaces in the world. So like our real world, in which we live under some rules, limitation, etc., the cyber space has found its rules, capabilities, limitations, trainings and instructions.

The relationship between the cyberspace and the real-space cannot be visualized only by indicating link between spaces. Synthetic space is the one in which we live in. This new image can help us understand the in-depth idea of the internet common culture. Everyday millions of people from around the world started changing different concepts such as identity and interpersonal interactions. Many of the changes have resulted because internet users agree to the relationship between real and cyberspaces. When the cyber and real space comes into interaction, cybernetic space comes into existence. The concept of spaces can be learned from the locations of human beings on the planet earth. This concept has encouraged human beings to explore the world and find new lands. It has also spearheaded the development of new technologies, which have rapidly changed our way of life. Spaces can be identified with the geographical location of a human being's home. New technologies are enabling people to locate the exact coordinates for any location. The use of sophisticated maps, satellites and other global navigation technology has assisted human beings in knowing their exact locations.

© Springer International Publishing AG, part of Springer Nature 2019
S. H. Sadeghi, *Pathology of Learning in Cyber Space*, Studies in Systems,
Decision and Control 156, https://doi.org/10.1007/978-3-319-91449-7_2

Capabilities and Limitations of Cyber Space

Capabilities

By the mid-1980s, personal computers became more and more commonplace with the availability of a large and growing number of people. Nowadays, people use PCs to communicating via the Internet in the broad global network, they enter into the cyber space, transfer information and etc. so personal computers are the most important tools having central role in communicating. Consequently, it can be said that human life is affected by newfound advancements in the field of technology alongside the high capabilities and Useful and harmful potentials of cyber space.

Besides providing teaching materials such as lecture notes, slides, streaming audio/video presentations, and extra materials, it can be used for setting up online quizzes, discussion groups, and uploading assignments. This is done in line with the present common electronic education technology, which assumes putting on the Web course materials that can be further fortified with elements of interaction such as discussion groups and quizzes. These capacities and capabilities of the cyberspace are to a great extent led to the emergence of new concepts and terms such as "electronic publishing", "electronic banking", "e-commerce", "electronic business", "electronic city", "education and E-learning" and the like. They also has managed to take many of the activities of the real space into the cyber space. Thus, one can conclude that the cyberspace can revolutionize all aspects of human life by transforming them from their present forming objective space into their new forming cyberspace. Some of these transformations might be desirable to choose and adopt, some might not, and we would not have to accept and even consider them. These cyber space's capabilities generally include:

1. Spread and variety
2. Maintenance and protection of information and its recovery on a wide scale
3. Ability to search in vast resources
4. E-mail
5. Extensive provision of e-services
6. Providing business-banking services
7. Ability of bilateral and multilateral interactions for users
8. The ability to create e-government on a large scale
9. E-learning
10. Creating an electronic city
11. Creating e-employment
12. Ability to inform and advertise in cyberspace
13. Ability to conduct a referendum and electoral campaign and promote the representatives
14. The ability to create e-health
15. Ability to create and develop scientific publications

16. Image capability
17. War and electronic dispute
18. The ability to produce second life
19. Virtual tourism, virtual tourism and virtual entertainment.

Limitations

We all depend on the capacities and capabilities of the cyber space day by day. But in fact, despite the great capacities and capabilities of this space, it has limitations, either for entry into this space or the space of this space. Accordingly, the general limitations of cyber space are:

- The lack of inadequacy of citizens' e-literacy to enter or use the cyber space
- The existence of technological inequalities between holders and pioneers in the field of information and communication technology will create a gap and create a constraint for other countries
- The cyber space limits the privacy of individuals, communities, and higher levels of government
- Trouble Racing
- Rising unemployment
- Challenging governments and traditional governments
- The limitation is due to the high costs for the infrastructure for creating this space
- Low security constraint
- Limitation due to lack of direct interaction
- Limitation due to non-connection to the Internet
- Limitation due to the interruption of the Internet
- Limitation due to the lack of infrastructure required to access the Internet
- Limitation of Supervision and Management
- Limitation of anti-ethical propaganda, racist propaganda, anti-religious propaganda and ...
- Limitation due to the intensification of one-way flow of information
- Limitation and difficulty resulting from inappropriate censorship
- Worms, viruses, hackers
- The limitation of the difficulty of finding the desired items and the high probability of deviations and obtaining unwanted content
- Slight constraint when transmitting pictures, videos, and sounds.

Rights and Jurisdiction in Cyberspace

Significant expansion of communication and information technology, especially the Internet, in recent years has made a fundamental difference in the lives and inter-actions of humans. Legal systems, with the acceptance of the fact that the law is not an indisputable principle and the imperfection of the law in practice and in relation to the issues raised, has addressed the challenges of this technology to modernize the laws and regulations and administrative regulations. This belief has been institutionalized by the legislator of developed countries, with whom the advancement of science and technology in different fields should prepare and elaborate the necessary rules for responding to problems such as the privacy of individuals in the processing and transfer of personal data, e-commerce, intellectual property rights, civil liability, and etc., In this regard, it is not surprising that, with the development of the Internet, the network is connected to the "real world" and legal reality has entered into virtual space; where words such as "rape in cyber-space", "cybercrime", "cyber terrorism", the cry for setting it is far off and unlikely. Different aspects of financial and commercial matters, such as tax laws, intellectual property rights, e-commerce, etc., have led governments to opt for the regulation of Internet rights. So, today, other governments cannot disregard the right to set up cyber space rights and assign it to their users or Internet service providers. But also various aspects of financial and commercial matters such as tax laws, intellectual property rights, commerce electronics, as well as good politics and ethics, the excitement of public opinion, and … have been forced Governments to take the lead in regulating cyber space. But given that the rights and rules of each society are basically meaningful within the boundaries of its borders, and also the fact that the cyberspace has no terrestrial boundaries, there are a large number of computer networks that produce information without any geographical barrier to the other side. This poses an important questions with regard to national boundary, which person or persons at national or international level have the power and authority to regulate this space?

At the moment, what named cyber rights is entitled to include all aspects of civil, commercial, criminal, international, and technical. Other terms have already been used and may still be used, which have been used in terms of the conceptual and historical scope, but have now given their place to this term. Cyber space right is one of the expressions that is equivalent to computer right. In specialized books, such as The Book of Jan Lloyd (IT Rights), there are some headlines that have been mentioned, "The Carin Tiger" (computer rights). The other term borrowed from the French word is an informal law, which is precisely the equivalent of computer rights. This is because in French the term "informatics" is translated into computer science. All these terms conceptually emerged in the first generation of IT rights and the term. In this regard, various efforts have been made in the form of treaties and protocols to shape the rights of cyberspace, the most important of which are as follows:

- **e-commerce**

Today, electronic communication technology plays a special role in commercial exchanges. The exchanges of individuals in the virtual environment are carried out in various forms and for different purposes. The Internet, which employs over a dozen billion people worldwide, has a special role in e-commerce, and the legal aspects of this type of trade are considered by legislators from different countries and internationally. As a result of information technology, electronic business-based business transactions have replaced traditional paper-based methods, and these transactions are entirely carried out in a virtual space without the physical presence of interlocutors. Extremely fast speed, low cost and access to other providers are a special feature of this environment and have created a global market based on the principle of competition and free trade. In response to the legal issues raised by this environment in the last decade of the twentieth century, legislators recognized the advanced countries of business in the virtual environment, spreading the same credit in the non-electronic environment. In this regard, legislative measures have been taken to allow the conclusion of electronic contracts and related issues such as security of electronic exchanges, electronic signatures and various types thereof, including scanned images, handwritten signatures, digital signatures, biometrics and electronic signatures certificates.

E-commerce has internationalized, and buying products online across national borders has become straightforward and convenient for consumers, providing new business opportunities for both domestic and international online stores (Hallikainen & Laukkanen, 2018).

E-commerce is conducted using a variety of applications, such as email, online catalogs and shopping carts, EDI, File Transfer Protocol, and web services. This includes business-to-business activities and outreach such as using email for unsolicited ads (usually viewed as spam) to consumers and other business prospects, as well as to send out e-newsletters to subscribers. More companies now try to entice consumers directly online, using tools such as digital coupons, social media marketing and targeted advertisements.

With regard to the role and importance of electronic exchanges in today's world, the American legislator has adopted a special digital signature law. Under the Digital Signing Act, any sign that has the purpose of certifying a post is deemed to be a signature and this signature is a commitment and obligation.

It creates for its owner. The intention and intention element are subject to the signature, and its registration is legally valid. The United States has reduced the scope of the requirements of the handwritten signature, and more than anything else, has put meaning and intent into its focus. The Digital Signing Act has therefore sought to reduce the barriers to electronic certification and remove uncertainties in this regard.

- **Privacy**

The privacy of individuals in electronic communications is one of the important issues of these days in different countries, especially in advanced countries.

Recognizing the privacy of individuals who have direct and close links with the personality and are considered to be the most fundamental human rights is strongly supported by the legal systems of the above-mentioned countries. In this regard, it should be acknowledged that, in addition to traditional methods commonly used by the ruling forces, modern technology has created a suitable basis for violating privacy by the public. New tools should be created in the form of special legal protections to protect privacy. The European Union has made a lot of precautions in this area, and by providing a comprehensive interpretation of the scope of the protection of individuals' privacy, it has committed member states to draft new laws and regulations concerning this right. Some of these members are the United Kingdom, France, Germany, and Ireland. In this laws and regulations the security of the common e-commerce, user, service, value added service and electronic communication services, confidentiality, processing, transfer traffic and spatial data have been carefully considered.

- **Intellectual Property Rights**

Intellectual property rights are important issues in cyberspace as well. Specific features of this environment include the elimination of physical and geographical boundaries, and easy access to the artistic, literary, scientific, and intellectual creativity of others. It is clear that the intellectual creations of individuals in the form of electronic data have intellectual property rights.

Intellectual property (IP) means all that human beings create with their intellect and intellect, such as art, technology, science, visual arts, and writing. Intellectual property rights are attributed to rights that are given to the inventor or the creator of work for a lifetime or for a period of time in order to protect his work or invention or idea. These rights are such that they grant a fully exclusive right to the inventor/ creator or his lawyer to use his work in any way for a while. It is clear that the intellectual property right plays an important and important role in today's economy. So they eventually came to the conclusion that the author of spiritual effect and his invention should be very important, since the invention of public goods comes from the same mindset (Saha & Bhattacharya, 2011).

As we know, intellectual property has a financial value and is economically viable. The circle of this property can cover all the instances and manifestations of non-imaginary and valuable human efforts. These rights are protected by international conventions and international organizations, including the World Trade Organization. In the process of accession to the World Trade Organization, the legislative actions of the applicant government are considered. These rights have been protected by all legal systems since decades ago. Intellectual property rights include copyrights, patents, and industrial and commercial property rights. In the cyberspace, various legal systems have tried to prevent violations of the relevant laws by including civil and criminal guarantees, because in a cyber-environment, it is possible to violate and override these rights far more than the real environment. Significant advances in this space require the reform and revision of laws and regulations on the distribution of these rights in many years, as legislators from different countries, including Britain and Germany, have been reforming and reviewing the laws of this area every decade.

- **Criminal law**

The cyber space has created new crimes that, in terms of the nature and manner of committing with types similar in the traditional form, they are completely different. Perhaps it can be claimed that in many cases the harmful effects of the offense are far more extensive than its traditional form in real space. Along with the advancement and development of e-communications technology, cybercrime is expanding. This has led, on the one hand, to legislators from different governments to reform and/or introduce new laws in this space, and, on the other hand, international organizations will also take important steps towards the establishment of criminal law in international arenas. Given the legal principle of "crime and punishment" (the current verdict or termination can be a crime that has been determined in the law for that punishment), lawmakers in advanced countries tried to enforce their criminal law in the cyberspace, given this unequivocal principle update. They determined the two principles of crime and its nature, according to the principle of the "proportionality of the punishment with the crime", and then, for the perpetrators, the punishment was commensurate with the type of crime.

Of course, there may be instances of cybercrime cases that are similar to traditional types of crimes, but it should be noted that the way of committing in this space is completely different. In addition, there is the possibility of committing crimes in a place where its material element is different from the normal types of crimes. The increase in the proportion of crimes caused by the formation of this space has led to division into this area to aim at the formation of the criminal and legislative policy of computer crime. Classifications include classical crime with cybercrime, crimes against content, crimes against information technology, Telecommunications crimes and cybercrime, each of which requires its own law. International treaties, including the Budapest Criminal Cybercrime Treaty and its addendum to Strasbourg, have taken an effective step in regulating this space.

Thieves and cybercriminals need less resource and less money to counteract, so the law can use methods to deal with them, which are somewhat different from real-world methods. The low cost of cybercrime is partly because computers and the internet have made the easier and less costly way for offenders. In these circumstances, the criminal law has the duty, in addition to punishing cybercriminals, to limit the low-cost and easy ways of committing a crime. For example, if the offender does his/her offense with computer, According to the law, the computer is identified as the successor to the offender, but the law must adopt the principles that the computer will be known as pseudo-successor and its responsibility becomes conditional. But some countries have imposed restrictions that have led to the loss of capacity and has prevented progress. Encryption, in addition to enhancing security in communications and increasing the freedom of individuals, has the potential to increase and growth of terrorism.

- **Civil liability**

Civil liability is one of the most important issues in cyberspace. Civil liability is the obligation to compensate the loss incurred by the lender, so when a person injures or damages another person and is obliged to compensate, he has civil responsibility. The fact is that the security, health, personality and property of individuals should be respected and in case of violation, the damages will be compensated. This responsibility is considered in two situations, that is, contractual liability and non-contractual liability. The contractual liability arises because of non-performance of obligations arising out of a contract or contract and non-contractual liability in the absence of the contract and the occurrence of damage to the other intentionally or by mistake.

In the cyberspace that different individuals are doing their interactions and, incidentally, the volume of these interactions is very high, the possibility of entering damage to the other person's influence is very high. Since, according to the general legal principles, no damage has to be compensated, this issue is of particular importance in this environment. In the cyberspace, support for users and subscribers should be addressed to service providers, agents, communications networks and trusted third parties. The user or the subscriber may suffer damages resulting from a failure by someone who has failed in his legal duties. In the past decade, claims have been filed by users and subscribers against service providers and trusted third parties in European countries, leading to the issuance of claims involving compensation by service providers. In the processing or transfer of personal data of users and subscribers of the rules and regulations of civil liability they must provide the conditions that individuals can compensate for the damage caused by the cybercrime. Hence, in order to realize civil responsibility in this environment, there must be a compensable loss, a harmful act and a causal relationship between harmful conduct and harm. Two examples of violations of the rights of others in cyberspace are:

1. Violation of rights through the establishment of links (links) between sites

One of the features of the Internet in comparison with other communication devices is the ability to connect different documents and elements to each other. These connections are more likely to be made by establishing links (links) between different Internet sites. Linking is done in two ways: When a user clicks on an icon or text that is capitalized, the link transfers the user to another location. Destination may be different from the same site or site, which is often the case. The latest type of link called "Link Out" is best known among Internet users. For example, the advertising section of a website that represents an air travel agency has links to increase the impact of advertising on its site to hotels or other places to use images and images related to the destination (Akbari, Moslehi, Fathi, & Bozorgmehr, 2007).

2. Infringement of the right through a technique called a mouse trap

Some marketing sites use a technique known as "trap mice" for marketing. The signing of this method is that users are forced to stay on that particular site upon entering that particular site, and their computer search tools (the mouse) will run out of service for a while and when the user tries to use the icons like Back or Close the site leaving or closing that window opens the window of the Web site and prevents it from doing so. This practice sometimes misleads consumers and throws them into irrelevant sites (Akbari et al., 2007).

Office and Cyberspace Cyber Space

With the arrival of electronic networks, and the influx of human activities and interactions into the cyber space, new questions have been raised about the implications for national sovereignty and how to manage and manage this space. The cyber space, which is the result of telecommunication-based computer networking, is designed not to be based on the logic of the national political frontiers, and thus the virtual world does not coincide with the political boundaries of the land between states-nations. This space has an open architecture and is basically redesigned. In this space there is a new boundary that consists of screens and passwords that separate the ones inside the network with those outside the network and thus separate the cyber space from the real world of particles and atoms. For example, there is no relationship between a network address with a network site and a territory of territorial sovereignty. The cyberspace has no boundaries on the land, because the cost and speed of the message transmission on the network is almost independent of physical location. No matter wherever you are, you can send your messages from any location and physical location to any location and your message does not face with any delays, physical impediment or deterioration. In this space, there are not any territorial or regional boundaries, breaking boundaries is a key paradigm for regulatory management. With the cyber space existence, Political and economic communities do not need to contact physically based on geographic proximity, they do not need to do much with cyber space. So, a study from governments showed that "so far, the biggest barrier to sharing is information, confidentiality and security, and the impossibility of administering and managing this space by governments", and this barrier is mainly manifested in the concerns of countries about the information that governments have about them. In sum, it can be said that cyberspace has no control panel because it is designed without a central control mechanism and is only created by a large number of computers and networks to which it is connected. But this does not mean that the cyber space is not controlled at all. Governments try to control and manage the cyberspace in different ways by filtering threatening and opposing sites and creating other constraints.

Accordingly, since the formation of the cyber space has created new risks, though the number of its capabilities is increasing day by day. Thus the management and control of cyber space deserves more attention. In examining the pros and cons of this space, policy-makers and decision-makers need to increase their awareness of management, e-security, and secure business confidence in cyberspace. As the e-security industry is growing and becoming more globalized, they need to dispose public policy challenges in the field of competitive politics, potential conflicts of interest, also certification. In the past, companies providing e-security services generally operated in three areas: access, use, and evaluation. In addition, today's industry includes companies that provide other services in this regard. Services such as data monitoring and filtering, aggressive reconnaissance services, fire walls, permeability tests for investigating the level of vulnerability of software and hardware, services or encryption software, identity authentication services or software by means of passwords, tags, keys which confirm the identity of the groups or the integrity of the data. Traditionally, the telecommunications industry was considered an essential element of welfare, comfort and public health, and therefore a core component of its development was public service. At the moment, access to cyberspace, through which e-commerce services in many countries are necessary for life. In developing public policies to create or modify the criteria for the management and administration of cyberspace, there are a number of important pillars:

- Appropriate legal and executive framework
- Technical and management measures to ensure electronic security
- Strong monitoring and prevention; to create better incentives for implementing appropriate and well-structured risk management systems, including e-security for financial service providers
- Digital signatures
- Sharing information
- Training citizens, staff and management on security issues.

Security in Cyberspace

Most people think of hackers and attackers when they talk about cyber security, and they assess security or insecurity in cyberspace on this basis. But with regard to security debate in the cyberspace it should be noted that the security concept is beyond the scope of this book. In fact, it covers any kind of challenge and damage to users, data, information, and so on. Accordingly, the US National Strategy defines cybersecurity as: "Cyber security is a large area, including the physical and virtual security of cables, computers and other Internet-related tools." Not only the technical view of cyberspace security should be taken into account, but its social point of view should be taken into consideration. Lots of damage is caused by social engineers through the cyber space. In this case, attackers are not the only system hackers. On the other hand, security in the field of information security is called management and technical solutions that ensure the confidentiality and integrity of information, access to information and the availability of information, services and systems by the service. Electronic applications are used. When you are safe in the cyberspace, access to information resources under your control means that no one can access these information sources without your permission. These resources include networking, processing, and personal information. Security in the cyberspace depends on the control of access to personal information resources, that is, no one can access such information sources without permission. On this basis, security in the cyber space will be established when a set of conditions is created. In other words, when this set of conditions is created, we pass from insecure to a more secure state. The goal of security in the area of information exchange is to preserve capital within this space. The capital of this space is data and information that is considered to be the capital of this community. Also, the systems that are being offered are one of the assets of this space to prevent them from being exploited. As we know, the security of this space is not limited to itself; naturally, security in this space can also affect the real world of society, and we see the real community affected by this space. In this regard, in the end, we must say that our material and spiritual capital in the real society is one of the goals we should not be afraid of. The space outside this space is the children of this country who use the computer every day and may be at risk from this space. In general, it cannot be said that this space is threatened, but through this space, the real space is threatened.

This introduction introduces the truth about the security that is absolutely impossible in real life and in cyberspace; however, security that is reasonably appropriate is achievable in almost all environments. There are various ways to use reinforcement mechanisms to increase and maintain security. We have physical mechanisms to ensure our security: high-rise buildings, tight and impenetrable doors with many locks and keys. It is possible to rely on other physical boundaries, such as walls and other barriers. We can also focus light on areas where they can penetrate. Finally, if necessary, we can use warning systems to detect intrusion factors and stronger protective forces, with the assumption that initial intrusive actions are not successful. Our familiarity with physical security measures is not as much about the nature of their cyber space, but understanding and how they can be used to provide security in the cyber space (such as the real world) is necessary, because both in the real world and in the cyberspace, we need to protect and defend ourselves against the attacks of others and, if the attacks are successful, we will compensate and reduce their harmful effects.

Cyber Police

In line with the Information Technology Revolution, law enforcement in the cyberspace encountered problems and complications from crimes and abuses. Due to lack of specialized forces to enforce laws and confront these crimes in the cyberspace, the amount of these crimes has increased in this space. This has led governments and countries to try to prevent these crimes. In this regard, countries began to train cyber-police specialist forces. So, with the formation of cyberspace and its development at various levels of society, future policemen will have to enforce the law not by guns but by their brains.

Social networks on mobiles also includes threats and dangers in addition to its outstanding services and achievements for everyone. Therefore, it is considered necessary to know the rules of how to work as well as how to deal with a threat in this area. Today, you can complete important cases without leaving your office. Bring out in the future, the police officers who resist the computer revolution will never step down. The idea of a police officer completing the case without leaving his office is unlikely, but a police lieutenant named Bill Baker is a real example of such an idea. Baker, working in Kentucky, with breaking up the success of a children's pornography band in Britain. Sussman, a journalist, says: "An e-mail from a source in Sweden led Baker to a website in Birmingham, England. After about 3 months of research involving downloading 60 pages of child pornography files and 400 images, Baker contacted the international police, Newcastle Yard and Birmingham police and managed to arrest the distributor of the photographs". So it seems that the police can deal with hackers in the computer field. In terms of the need and necessity of the existence of cybercrime, cyberspace is a shadow of real space in the electronic environment, with all its capabilities and capacities, and can directly and indirectly affect the real environment. And it is important to recognize that the existence of cybercrime is essential for controlling and preventing cybercrime.

Cyber Terrorism

The term "cyber terrorism" is growing in popular culture, but it is still difficult to define it. While there are some popular and extensive definitions of it, the exact meaning is not specified. In general, cyberterrorism means using computer resources to threaten and force others. For example, hacking a computer system of hospital and change the drug prescription of a deadly person for revenge (Gordon & Ford, 2002). In other words, beyond the cybercrime, the Internet enables the political groups and other groups to gain an opportunity for anonymous use to attack the vulnerabilities of the national states. According to US law, cyber-terrorism is defined as:

The perpetrators behind these conjectural attacks were often called "cyberterrorists," a term whose provenance dates to the same period. In popular accounts, cyber-terrorists referred to computer hackers who might cause airplanes to fly into each other, bring down the nation's banking system, or use computers to kill (Kenney, 2015). A deliberate political movement against whatever connects to cyber space including information, computer systems, computer programs and data that may be a precondition to violence against civilian targets and creates by sub-groups or by agents. The goals of the cyber terrorist are simple. First, they may attack data or control systems. Second, since the cyber-terrorism goals are damaging important things, this leads to severe damage or violence towards the country's main interests. On the other hand, Walter Lacourt, a terrorist specialist in strategic and international studies, points out that a CIA official has claimed that he can "cripple the United States with $ 1 billion and 20 hackers." Lacur reminds, although the target of terrorists is usually the murder of political leaders, hostage-takers, or sometimes a sudden attack on public or public facilities, the damage that may be caused by an electronic attack on computer networks can be "much more sad and its effects, to stay". Lacourt believes that computer terrorism may be more devastating to a large number of people than biological or chemical wars.

Accordingly, cyber-terrorists, like "ordinary" terrorists, have political motives for committing crimes. However, their risk is not limited to government computers private corporations and even non-governmental groups are also vulnerable to the harm. This is because depending on the specific political goals pursued by the terrorist groups, they may attack any computer that contains the information they are seeking. There is substantial research trying to deliver exact definition of cyber terrorism and investigating to qualify and differentiate the exact nature of cyber terrorism as against hacking, cyber-squatting, hacktivism etc. and strategies to counter acts of cyber terrorism and cyber warfare (CWCT). CWCT have several kind of strategies supported by intelligence on the spectrum of APT (advanced persistent threat) that the research findings show that decision makers and policy makers must have a sensitive plan to make readiness for capabilities for cyber terrorist's attacks and its effects (Bagchi & Bandyopadhyay, 2018).

War in the Cyber Space

First of all, in order to understand the concept of war in cyberspace, in general, we need to distinguish between two types of warfare in the cyber space. These two are:

- A network war is an information warfare that flows between individuals, communities and nations, or directed toward the community and civilian purposes. The purpose of this war is to destroy and suppress existing thoughts and ideas in the community and replace the new ones.
  Network warfare is targeted in two ways: one public opinion and another one of the elite beliefs. This war involves a wide range of diplomacy, propaganda, psychological maneuvering and deception using local media and infiltrating computer networks and databases. Disruptions and manipulations in the computer network and information systems are in this context.
- The cyber type of this war is aimed at breaking into the information and communication systems, control and command systems, communications, intelligence and intelligence of enemy forces and others.
  Their operation is carried out on the battlefield or in the normal state. In other words, war in cyberspace can be a reflection of both "cyberwar" and "network warfare". Network warfare is interpreted as an advanced electronic warfare that refers to different cyber domains; in the other direction, cyber warfare is precisely linked to a military breach, C4I (command, control, communications,

computers, and military information), and recently C4ISR (command, control, communications, computers, military information, monitoring and identification) or C4I/BM (command, control, communications, computers, military information, war management). But in general, according to Arquilla and Ronflet's main paper, war in cyberspace is guided and geared for leadership, military operations based on information principles, an attempt to know everything about the enemy as long as it takes the enemy away from the necessary knowledge. Holding and interrupting and/or destroying the hostile information and communication systems.

According to usage rate reduction of some older technologies such as TV, radio, telephone and etc. and taking their place by internet, mobile software and computer applications, we can't say technology is just rising, but it is moving the borders between reality and conceptual. So definitely next generation of wars will not be same as before (Adams, 2017).

On this basis, generally speaking, network warfare against the community is taking place and the cyber war takes place against the military forces. Some harmful activities derived from human errors and faulty codes but more important alarm is different attacks on various kind of devices through special destructive codes such as "viruses", "Trojan", "horses", "spywares", and "adware" and etc. (Fadel et al., 2017).

The war in cyberspace is done by certain tools, some of which include:

Viruses: Viruses are programs that can replicate themselves in programs and they are activated when the host program starts to operate, and subsequently replicates the virus and infects other programs. Viruses are built in cyber space. So it's not surprising to be used as an intelligence weapon. In that case, the target network fails, or at least it creates a wide range of inadequacies. Actually virus is a kind of malware that enters into a computer in a variety of ways, such as an email or a file infected with it, which puts its code in other computer programs, changes them and duplicates itself. When the virus successfully completes its duplication, it is said that the computer or affected areas have been infected. For example, a kind of famous email-aware viruses that could infect around 20% of computers by special method of infiltration called "Melissa" caused $80 million scathe all over the world. This virus would resend to several contact of receiver and this way it could disable mail servers by huge flood of emails transferring between users. David Smith who made this virus condemned to jail for 10 years (Chakraborty, 2017; Zhu, 2016).

Worms: Worms are a standalone program that flutters itself from one computer to another, often on the network, and does not change other programs unlike viruses. The destructive messages of this weapon can be examined in two ways: one is the destruction of information resources in the network and the other is the transformation and dissemination in the network. Nowadays worms are detecting fast by antiviruses but on the other hand high accessibility of internet and file sharing made worm extension easier like the famous one that known as "ILOVEYOU" (Broadhurst, 2017).

Trojans: Trojan is a malicious computer malware that affects a computer by misleading users about its real aim and contrasting viruses or worms, the Trojan acts indirectly and hidden, which can get some sensitive OS access without permission and then attackers break the privacy, authentications, passwords and other private data. Trojans aim to both, home users and enterprise level organizations (Li, 2017).

For example Gingermaster is a kind of Trojan targeted Android operation system that installs hidden malware while user is installing other applications (Zhu, 2016). The GingerMaster malware conceals some packages of codes to access system files and folder that foist this package instead of installation package with the same name as install package (Suarez-Tangil et al., 2017).

Troy Horse: Troy horses are programs that hide within other programs and run their own program. Troy's horse can camouflage itself, even in network security plans.

Bots: Another type of malwares are Bots and Botnets which let hackers to take control over a pretentious systems. They call affected computers as "zombies" that can be control through internet connection. While a network of many systems be affected by botnets, that network can publish viruses, produce spam via internet by different ways, and obligate any kind of cyber-crimes. Actually a bot can be any kind of known malware. Bot can change itself to a worm or act as a spyware like key-loggers and password collectors, packet monitoring or record credit accounts (Zhu, 2016).

Other cyber warfare tools include logic bombs, backbones or livestock; chips destruction; germs; electronic disturbances; EMP bombs and HERF guns; electromagnetic pulses and any other kind of these tools grow for a period of time and some of them fade forever. Accordingly developers should upgrade and improve their skills at the same time as hackers to prevent huge network destructions and keep safety for information.

Chapter 3
Training in Cyberspace

Introduction

The process of education, which addresses the fertility and flourishing of thoughts, is one of the important things in human social life and nowadays it's getting more importance at the same time as improvement of technology and modern lifestyle. In this way, educational issues gradually increased and need serious attention in social or individual aspects. Therefore, parts of the resources are devoted to this matter and, of course, its contribution to the increase can be paid attention. Nevertheless, educational needs are diverse and widely accessible, and they are met in different ways. In this regard, it is obvious that the advancement of information and communication technology and the formation of cyberspace will also lead to their contribution in education and learning. Therefore, given the capabilities and capacity of web, these advancements will make the web a powerful remote educational media in order to better accomplish the objectives of the education.

Cyber education have challenged providers, students and teachers from two technological and pedagogical dimensions. Working with cyberspace is so that newcomers and experienced computer users should be able to use the internet and virtual tools as the skilled ones quickly, but on the other hand, these people have to work with the Internet and cyberspace so that they can handle the challenging and obscure concepts and the core content of technology and manage procedure of the organization, transfer and integration of information, skills of movement. One of the disappointments is that technology has made it hard to comment on students' documentation and assignments, but it should also be noted that using technology we can digitally store student assignments and writing process and at the same time, we can publish the work of students online (Mkrttchian et al., 2018).

© Springer International Publishing AG, part of Springer Nature 2019
S. H. Sadeghi, *Pathology of Learning in Cyber Space*, Studies in Systems,
Decision and Control 156, https://doi.org/10.1007/978-3-319-91449-7_3

Types of Training in Cyberspace

The top of this space, as well as interesting designs and emergence of various forms that have learning-oriented, interesting, interactive, efficient, flexible, meaningful and facilitated features, etc., provide a ground for the improvement and development of cyber-learning education and learning. These changes and changes caused by the cyber space in the system of education and learning have brought about an evolution that necessarily results in its own culture and has appeared in various forms. On the other hand, high capacities and cyber-space capabilities have created opportunities for creating new learning environments for governments, organizations, formal institutions, informal institutions, groups, and even individuals. Despite the changes in cyberspace, the forms of education in this space are still hidden in different ways from the forms of education and learning in real space (namely formal education or informal education), but using the possibilities and resources in the cyberspace are trying to educate users and students.

There are several methods for e-learning that the most important ones are as the following (Xu, Yang, & Zhong, 2017):

Face:

In this method, learners and learners online are physically present on the class, with the exception that in these classes, e-tools are also used, such as the classes in which PowerPoint is used to convey content.

In this method, the relationship is between the people in the classroom or that is done through the telephone, for this reason, there is no electronic communication in this way.

Non-simultaneous:

When content collector recovers content and learns content, there is a delay between transferring and accessing content.

In this way, people can interact with each other through various tools and technologies. In this method, technologies such as e-mail are also used.

Tutorial:

This kind of learning that is a virtual learning method is the same as self-reading. In this type of learning, information is provided in formats such as CD or DVD.

Combination:

This kind of learning is also a combination, but with the difference that it combines face-to-face approaching. In this method, learners and learners meet in both face-to-face classes and virtual training.

Training Resources in Cyberspace

Considering the widespread dimensions of human life, allocation of resources in various fields is carried out and education is one of the most important of them, that from the beginning of human life as a basic and urgent growth of need has been associated with him during different periods; in order to meet this need, various educational resources have played a role in formal, informal and hidden forms. These resources were from the family, society, schools, the media, etc., which, through training in various forms, tried to create and facilitate learning with their own goals and society among learners. In the course of technological and information developments, these resources have been more and the impact of some on education has decreased. In this regard, with the formation of cyberspace and the entry of individuals into this space, there are many educational and learning resources based on the cyber space architecture. These resources have been created through cybercrime using a variety of programming for the implementation of business logic, service provision, e-learning, and has created new educational resources for education. Also, the learning of people has taken into account the needs and advancements of the communities. These new teaching resources, which originate from real-world sources, and in a new form of education for cyber-space users have a variety of types that are presented below the most important ones based on formal, informal, and hidden cyberspace education, have been:

Official Sources of Training

Academic Centers and Training

Information and communication technology has made comprehensive developments in various military, political, economic, social, cultural and educational fields. Although the beginning of its transformation was from the military, largely by maintaining military secrets and preventing the loss of information sources as a backbone of military power, it has gradually been declined due to its profound influence on scientific and academic environments and have greatly affected the rate of universities and scientific environments, so that in the short term, many developed countries have made use of this opportunity. Of course, it should be noted that in most parts of the world, the most effective leap forward has been to apply ICT in higher education since 1990 (Stensaker, Maassen, Borgan, Oftebro, & Karseth, 2007). This trend has been such that today ICT is considered as the axis of development in universities and research centers; as long as the establishment and implementation of a cybernetic university has been accepted among the experts and the scope of influence and expansion. It has been developed not only in developed countries, but also in developing countries.

University are using communities to reach united learning methods and e-learning systems. So students experience a learning system derived from their needs (Sadeghi, 2018). The essential subject to show amount of e-universities' usefulness is to explore clear reasons of innovative pedagogical practices (Bouchrika, Harrati, Mahfouf, & Gasmallah, 2018). In this regard, educational institutions, based on developments in information and communication technologies and the formation of cyberspace, have begun moving toward entering cyberspace to launch their university-based and distance learning courses, and in order to design their own courses. They have identified the learner's needs and the level of their specialized knowledge (Anderson, 2004). So that, now estimated, more than two-thirds of higher education offer e-learning courses, most of which are entirely on-line in cyberspace. In 2007, the number of students of electronic learning courses was more than 3.5 million, which is steadily increasing with an annual growth rate of 21.5% (Wu & Wu, 2008). Advantages of increasing the use of cyberspace in the global education system include:

1. Increasing communication channels through post-electronics, wake-up groups, etc.
2. There is plenty of flexibility in when and where teachers and students perform their duties
3. Access to unlimited amounts of resources and information
4. Bilateral communication at the lowest cost
5. Extensive use of multimedia software
6. Access to international information (often in English)
7. Continuous updating
8. Having a digital format and the possibility of applying changes and applying for other purposes
9. Having an overarching pattern
10. Encouraging links with other sources and the relationship between beliefs
11. Freedom of Information
12. Formation of virtual classes and virtual meetings

And ... (Gray, 2004).

These features and features of the cyber space have caused, in addition to universities, lower levels to try to use and exploit the cyberspace to educate and develop the scientific community and students. As the viewpoint of e-learning and education as one of the most popular and, at the same time, most refined fields of exploitation of facilities and capacities cyber space has become an optimal learning and learning resource and many books and articles in this area are being written in this field to justify the use of this space in teaching and learning in different parts of the world.

Governmental Centers and Training

In line with developments in the field of information and communication and the formation of a cyberspace with high capacities and facilities, different sectors of society in various fields to this space and has provided good groundwork for providing business, health, management, and so on. One of the parts that has entered into the cyberspace with a growing trend is the entry and formation of e-government and government organizations in this space. The government, which is the most important organization in a country and a management body, is responsible for its responsibilities to society and citizens, and ... tries to exploit the capabilities and capabilities of the cyber space in this field and create sites, blogs, portals, etc. in this space to provide services and achieve their desired goals.

In addition to providing government-provided services over the Internet, they are implementing projects for the creation of web pages in cyberspace that ordinary citizens can participate in order to establish a democratic society. The cyberspace allows citizens to raise their concerns and issues in this way with governments and help the government to better serve the country and meet its needs. One of these projects that is conducive to the democratization of society through which governments are conducting electoral and democratic elections throughout the country through the use of cyber-space facilities (Lee, 2006: 2–4). On the other hand, governments as the most important administrative and executive body in the country to achieve their desired goals in the cyberspace, countering the enemy's goals against the country in the cyberspace and ... need to promote and educate users at various local levels. National, regional and global in the cyberspace, in line with their goals, believes, and views. Accordingly, governments in different ways, in order to achieve their goals, confront the provocations of the enemies, express their views and ... need to provide training and learning in cyberspace, which is done in various web sites, government support websites, news sites, and more. Of course, it should be noted that given the dependence of societies and governments on the cyberspace, having the ability to provide education and learning in cyberspace has become a vital issue for governments and countries, and neglecting it can be the basis for challenges for them.

Private Centers and Training

As in the real area of private centers, they are active in the education sector along with the academic and public sectors. In the cyberspace, the private sector is also active in the field of education. The motivation of the private sector for learning in the cyberspace originates from the real environment, which in general these motivations can be expressed as follows:

(A) Economic motivations: The most important motivation of the private sector for learning, whether in the real or in the cyberspace, is from economic incentives. This motive has led to the private sector investing in cyberspace to provide virtual training courses, offering various services in cyberspace, advertising for companies and organizations in various ways to earn income, and so on.

(B) Scientific and developmental motives: The private sector feels responsible as a component of society for individuals at different levels of the society and, as it strives to expand scientific and developmental education in different fields for the learning of individuals in the realm of space. Cyber also does this in a variety of ways to improve and expand the scientific and developmental situation of individuals in the form of providing software training programs, providing educational texts, etc.

Among the training and learning motivations of the private sector in the cyberspace, one can mention the sense of responsibility towards the community, political goals, propaganda for a sponsor in the educational environment, and so forth.

Informal Sources of Education

Web Sites and Training

The web is a new and amazing technology for human life in the present era, which is used to transform the educational method in sectors and for various purposes. At the same time, perhaps one of the reasons for this plurality of perspectives can be seen in the progressive development of "technology-based pedagogy" and the new conceptual learning phenomenon in each step of this development (Dean, 2006). However, Web-based pedagogy includes all tutorials, contents and practices that are provided through the Web. In other words, web-based pedagogy includes any type of training provided through the Internet and intranet. In this type of web training, materials are provided by the Web server and via the Internet. This form of education, along with the benefits of computer-based education, has also provided the benefits of face to face traditional education in some areas.

Creativity can be defined through a new viewpoint as a product. So it should be organized and also consume if available with e-courses for many shows such as e-music, festivals, video effects and all variation of visual arts (Stoltz, Weger, & da Veiga, 2017).

The web is the most exciting and the best of the internet. Generally, the Web is described as a mass media cloud that retrieves information and aims to make more people have access to evidence globally. The web provides reliable tools for Internet users so they can access accessible resources on the Internet (such as images, text, data, audio, and shapes). Each organization and every web user can create a dedicated web page that includes any required information. Cloud text can facilitate interaction and communication between a proprietary web page and other proprietary web pages and this is considered as an advantages of cloud text. With these capacities and capabilities, the Web-based learning environments make it possible, firstly, that people are always learning and teaching, and secondly, with the elimination of time and space in learning, learning and teaching in any place becomes possible for him. This is not even limited to formal education courses, but technology will enable people to have the opportunity to learn for a lifetime and even anticipate that the age and standard education curriculum will find another shape.

Weblogs and Tutorials

The word blog is a mixture of words and blogs. The term "word" is a word in English belonging to the medieval period and is of unknown origin, and it means the "Office of the Report of the Cruise Ship". The word log in the computer specialized language is referred to as a file that records the events occurring on the computer. The word blog is also used in English in the form of a blog summary. Blog also has some synonyms such as weblogs, Weblogs or blogs which are pages of an internet web site as virtual places where people can write their favorite concepts and contents which are organized for certain purposes every time they want and arrange them from new to old, the weblogs may include many virtual information and other people can use them for different purposes such as educational, commercial, scientific, political, propagandistic, cultural, linguistic, etc. (Agarwal, 2009). In this regard, considering that blogs have high capacity and ability to deliver educational and informational goals, and have low costs, these types of pages are growing in the cyberspace.

On this basis, blogs are one of the most widely used publishing tools for web-based purposes that can provide written content as quickly as possible for public use on the web and as one of the interactive ways of communication in education and learning to be used. Educational and learning functions of the blogs include access to more audiences, quick feedback, thought protection, permanent access, writing, saving and retrieving knowledge and information, avoiding duplication, collecting the most important and up-to-date resources in one place and updating knowledge and information. Also, blogs can create a variety of advancements in educational environments. Blogs can be an important field for educational technology innovations, because they promote literacy through story-telling, and provide an opportunity for collaborative learning, every moment and wherever possible to provide the information they need and also can provide this training in all areas and goals.

In recent years, the effects of technology in education and training have been highly regarded by governments. One of the concerns of instructors is to increase student interest and participation in the classroom. One of the important effects of technology is to enhance students' motivation and self-esteem and increase team collaboration. Hence, the role of teachers and instructors in attracting the students' participation and interest is very high and their awareness as much as possible is required. The most important result of e-learning with instruments such as weblogs can be mentioned as the following (Özdemir & Aydın, 2015).

- Making Change in the Role of Students and Teachers
- Increasing motivation and confidence
- Technology skills
- Working more together with classmates.

Wiki and Training

I chose Wikipedia as a quick synonym alternative to prevent it from being called fast.

The wiki in the language of Hawaiian people means "Beginning" or "Oral" and is defined in Persian a collaborative site. It is referred to as "Web sites" that allows all its visitors, sometimes even without the need to register on the site, to edit it, i.e., adding or deleting content relating to a topic. The Wikipedia programs that make such sites possible are called Wikis. The first Wikipedia created on the Internet was created in 1995 by Cunningham and named Wikinews. The Wikinews word used in the name of this site was inspired by the name of one of the bus lines in Hawaii. Wikis allows users to create web pages about various topics without programming knowledge. In simple words, it's a web page that any person can correct. The initial view of the formation of Wikis is that any user of the Wide Web can now both read and write both the content and the Web site at the same time, so the Web correction process is simplified. For this purpose, the Wikis use simpler conventions and rules to refine the appearance of the texts laid down in Wiki, and these rules are different from each other in each Wiki. Another benefit of the wikis is that any changes to the pages are recorded, and the pages can be returned to the state before the change. In addition, there are wikis allow user to compare the content of a page before and after using multiple variations. On this basis, the Wikipedia provides the context for commenting on the topics and purposes of a Wikia, and causes users and users of Wikipedia to participate in teaching and learning with pre-designed goals. People participate in this space without any attention to the idea that this type of web-based learning is inducing users to learn and receive their views and beliefs. Because this environment provides active participation among users, they provide effective learning and learning.

Portals and Training

The word portal is one of the most useful words that has been developed in recent years. When we encounter changes in the information environment when studying, researching and learning. Unfortunately, in spite of the different definitions and perspectives for the word portal, there is no agreement or understanding on the specific definition of this term. In fact, there are a lot of obscure and confusing definitions for this word. Some people believe that only the word "portal" on the home page will make it a portal. After that, with specific links, especially the link to the search engine that gives your home page access to a number of web sites, this will be portal. Do you think this is a portal?

For the sake of clarity, it is best to divide the portals into two groups: (1) Horizontal portal, also known as Mega Portal and (2) vertical portals. Then we examine these two groups.

The horizontal portal is a public Web site that tries to provide users with all the services they need. Including news, weather, shopping, chat group, and so on. They force you to put them on your first page when you use the web.

The portal provides specific organizational information to the user. The University Portal should provide all the information that the portal provides. The difference between the horizontal and vertical portal is that the user must have a specific identity in the horizontal portal. So that when a user logs into the vertical portal page a specific page appears with his personal profile (Strauss, 2002). In this regard, it should be noted that in the domain of the Internet, the definition of the portal has retained its variable nature. In the simplest sense, the portal is the gateway to the web, the platform which each user must use as a source page before moving on to other destinations. Marshak, vice president at Patricia Sybold, says: "Portals reduce the amount of people roaming the web." At the same time, the portals are gradually becoming sites that each user uses throughout their day to manage a wide range of their daily activities, such as a complete review of news stories, stock prices, e-mail, and long-range messages. And joining the halls. Hence, portals carry out very different tasks and they all consist of five main elements: search, content, creation of groups and individual creativity programs. Organizational portals provide a set of reliable information paths by checking their users' information needs. Since the provision of portal services is essentially network-based, and in particular the Internet, service-oriented organizational portals perform beyond information addressing, which usually includes such things as search, e-mail services, polls, newsgroups and discussions, and so on. Interaction between the user and the portal's main website provides the opportunity for exploitation in the field of e-business. Accordingly, the portals have high capacity and capability. These pages have specific priorities and links that are specific to their intended purpose and structure, which asks the user to or unwittingly asso-ciates them, and provides them with an opportunity to create a learning and learning environment for them. Portals also have multiple capacities in the field where a single page can provide multiple facilities. These capacities allows them to attract many users and thus can directly provide images, texts and, etc. on their portal page. Provide training ground. On the other hand, by creating multiple links that are consistent with the goals and viewpoints of the portal, it provides the context for learning in its Web space.

E-mail, Messengers and Training

E-mail is one of the most popular and most commonly used tools in cyberspace. This communication tool is easy to use and has the ability to spread around the world. Individuals can use this e-mail at their home or workplace, through which they can use their customers, employees, students, and others in a broad and flexible way to educate their goals and perspectives. The e-mail feature is costly since other cyber space training resources must be made through a connection to the Internet.

On the other hand, the email allow the user to be online or offline and the outside of the interaction line, and then, after connecting in a limited time, sent file can be taken from e-mail and studied outside the cyber space.

Whereas E-mail took its place as the most formal tools to communicate between all groups of society, Messengers are growing rapidly for informal communications. Opposite of emails that sender has to wait for a longer time to get your answer; each person can contact others immediately. On the other hand, fast developing rate of networks facilities such as 4G bands and other easy accessible networks, messengers got high priority on human life. Messages are used as a normal text or other type of Medias to exchange information and daily conversations. In this method, instead of sending messages or other information through e-mail services and crowded inbox, they are sent to the mobiles easily.

Both E-mail and messengers can send contents to the receiver in different formats (text, image, sound, or video) and for different purposes, and provide a learning background.

Communication on this platforms is twofold:

1. Person-to-person relationship
2. Individual communication with the group through the postal list. Each person can create many accounts with different addresses and this addresses that have their extensions referring to a country or public domain, which is usually difficult to locate.

E-mail also has another powerful tool. This is a list of posts or list service. Using the Client List, individuals around the world can communicate with one another. After the advent of e-mail, users soon found that the ability to send an e-mail to a group of people was very useful for cooperation, training and negotiation. Therefore, the first list of services was created by the Information Center. The broker's e-mail list manages a large number of mailing lists, so that each specific area addresses the interest of network users, and each has independent members. The services, Medias, ideas and information are easily exchanged between users. These capacities and capabilities of e-mail can be a good and strong field for a useful, effective and interesting learning for users in order to meet the desired goals (Fig. 3.1).

Virtual Society (Social Networks) and Education

Originally, Ringold introduced the concept of virtual society. To define a virtual community, it is better to first define the term Net. Net is an unofficial term for interconnected computer networks that use CMC technology to link people around the world. Ringgold defines virtual societies as such. Virtual societies are social gatherings that appear on Net. These virtual societies involve the formation of networks of personal and social relationships in cyberspace (www.rheingold.com).

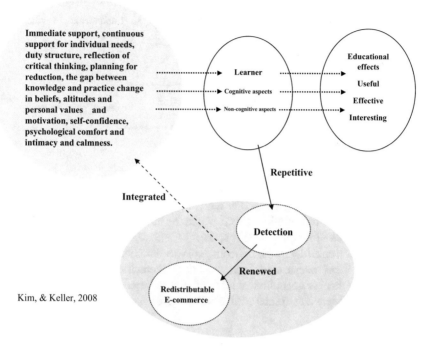

Fig. 3.1 The conceptual from work for using email in useful teaching

The virtual community can also be described as: "A group of people who may meet each other face-to-face or with one another, and turn their words and ideas into bulletin boards and networks". In this regard, the members consider the impact of communication culture, goals, views, and beliefs of these groups and social networks on themselves. Also, the identity of communication and views and beliefs (style and identity of the general actions of the individual in relation to others) may undergo some change to some extent in order to be considered as the member of group. On the other hand, the social network and the group in question also change and align the views and beliefs of the members in line with their goals. One person trains his goals to members in the framework of group formation, by participating in group discussions, sending emails, sharing pictures, and so on. In another word, it is the community of society that exists at least in the virtual world and the social group in question, in different ways, and it changes the attitudes and beliefs of the members in line with their goals and align them with their goals in the framework of the formation of the group, through in-group discussions, sending emails, sharing pictures and so on. However, this change in identity and beliefs may be limited to the time and place in the presence of these virtual networks in a cyber space. But there is no doubt that in the real identity of the members, their beliefs will not be ineffective. In general, all the components of a social network that a person interacts with in a cyber space interact with a person's unconscious mind and together act as

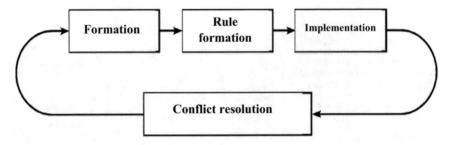

Fig. 3.2 The stages of the formation of the virtual group

a learning environment, and lead to the formation of a series of beliefs and specific views that is aimed at in the social network. Accordingly, these types of virtual communities and cybernetic networks can be considered as one of the most effective sources of learning in this space.

Researchers consider Social Media tools as influent leaders of teaching and learning practices because of their sincerity, interactivity and friendliness. But unfortunately there is not sufficient international surveys about actual usage of them in e-learning (Manca & Ranieri, 2016) (Fig. 3.2).

Messengers and Tutoring

Each of the Internet technologies in the cyber space has special capabilities. One of them is the chatroom environment (Fig. 3.3), which has a great ability to transfer audio and video, text, and more. Chatting is an English word for conversation. Based on this, chat rooms are considered to be places where people can talk together. Chatroom is actually an unstructured tool for chatting with people who are online at the same time. In chatrooms, people who have the same overall interest talk together in the form of a group so called they entered chat. These high capacity and forum capabilities can provide an educational environment among those who are in the chat room and provide a platform for training in different ways. They also develop a specific culture among users, and the thoughts, perspectives and the user's beliefs have transformed this environment and spread a specific culture among them. There is also the ability to chat in the context of a targeted topic between a specific groups and develop user feedback and responses, based on the program. Take advantage of their interest and provide their training in this envi-ronment and or they can create a space for users' questions and responses, and encourage users to pursue their goals and perspectives. On the other hand, this chat room has the ability to turn into rooms for formal education through education, higher education, and various organizations. But in general, the possibilities and capacities of chat rooms have led various social classes from different ages to find special tendencies for web chat, and many chat rooms are welcomed by many. The

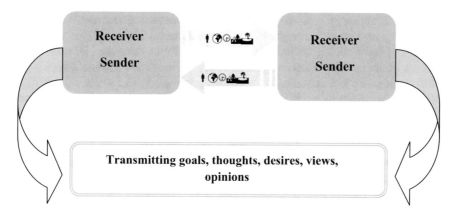

Fig. 3.3 Transferring goals and thoughts in the form of images

cyberspace of the dialogue, the anonymity of the people and the lack of social coercion in observance ethics and rules have created the virtual environment of conversation as a place for the development and promotion of thoughts, and hidden resource training.

Games and Training

In the modern world, video games have become one of the most important activities of children in their leisure time. In fact, today the video game industry has surpassed the film industry in some cases, and some TV officials now believe that video games have affected television audiences. Although there are similarities between video games and other media such as the network, the press, the movie and the theater, we must admit that video games are different. The interactive aspect of multimedia and games is that even when the books were the most prominent media, it was not seen, and this feature completely distinguished them. Games can also be informative and can be used for education and entertainment. In recent years, there has been a renewed desire to use educational games. The huge number of computer games in the industry has led all of us to ask this question what is the reason for this success and reputation? For this reason, researchers are trying to justify and undertake substantial research on the significant and inherent training that is obtained in most of these games (Egenfeldt-Nielsen, 2005; Prensky, 2006; Squire & Jenkins, 2003). It's obvious that all games are designed according to specific goals and beliefs, and they educate users about the goals and provide the context for learning and receiving the messages and goals desired by the game constructor. Of course, it's worth noting that the educational aspect of the games is not new, as in 1961, Kylewis claims:

If the game is to be played, it will not play.

In the meantime, the cyberspace, with its high capacities and capabilities, provides many backgrounds for games and entertainment. Children play cyber games online and enjoy surfing on the Web. According to this, there are kinds of tutorial and the goals that have been designed in format of these hobbies and games. Most of its excitement and entertaining categories teach users who are mostly children and teens. It should be mentioned that since the presence of an individual in these games is active, this type of training effectively impacts on the individual, and he well educates the goals, beliefs, and perspectives embedded in the games. For example, some of the games that are designed by the enemy and hiding behind the national symbols and religious symbols of a country can be pointed out. Or games dedicated to these issues in which targets were taken to hide their privacy and attack traditional families in countries.

Images and Training

The images contain special fields and content that provides different kinds of meanings to the viewer. These images, based on the location, goals and views behind them, have many things to say that come in the form of a video, a photo, a cartoon, a design, and so on.

The Advantage of Visual Thinking in Electronic Learning

1. Simplifying complex concepts.
2. Encourages online learners to organize key thinking.
3. Promotes learning and knowledge preservation.

The images have the ability to induce certain goals and beliefs in the very short time to the other. The value of the images is so great that it is said:

An image is worth more than a thousand words

In this regard, the cyber space with high capacity for storing images on one side, and the convenience and low cost of adding images to this space, provide the background and opportunity for individuals, groups, organizations, governments, and so on. By presenting their images in this space, they will in different ways train their goals and perspectives in this space, and they will be taught them in order to achieve the goals and desires of the case. Take advantage of the comment. Based on this, the images sent by the transmitters to the recipients in the cyberspace provide some kind of goals, values, and beliefs that the sender sends to the recipient in a very fast and hidden way.

Other Hidden Teaching Resources

Among other hidden caching resources in cyberspace are maps, cartoons, videos and more.

Overview

Today, advances in information and communication technology and the formation of cyberspace have had a tremendous impact on many human activities, one of the most important of which is learning and education. The main sources of human learning in the real world are the family, schools, schools, the media, and more. But in the new space, the main sources of user education include websites, blogs, portals, and wikis, along with games, films, pictures. All these have formed a large, flexible community to create and facilitate user learning.

Educational practices in this new environment, as in the traditional way, are formal, informal, and concealed. The learning process of users, such as learning in the real environment, involves two forms of intentional learning (scheduled) and random learning.

But the important point is that the emergence of this new space, in addition to providing diverse educational resources to governments, organizations, institutions, groups and individuals, has also led directly and indirectly its culture as well as their users and audiences and influenced their different dimensions of life.

Chapter 4
Learning Objectives in Cyberspace

Introduction

Any conscious, rational, and productive activity is determined by its intentions and objectives and it pursues certain goals. Education as an activity also has specific goals and the pursuit of education in any discipline or course is associated with some intents and purposes and it is carried out in order to meet exact needs and achieve specific targets. These educational goals express the desirable situation of students or people under training in an educational process. Educational goals are different in terms of the nature and time of realization; some are idealistic and poorer, a group of them can define the final results of an educational activity, and others specify the type of activity in pedagogy. In other words, the purpose of the education is to instruct perfectly learners about their educational activities.

New and modern forms of electronic learning have created opportunities that do not have any time and space restrictions, and you can use these educational facilities at any place and at any time. In order to improve the educational efficiency of this space, you can combine attractive exercises and methods with legal contents and use them in quasi-legal educational environments. The contradiction between the virtual and actual or traditional educational classes is untenable, and perhaps this conflict will remain forever. Today, countless people use mobile phones and computers and this has provided them with new forms of language learning and several types of language training. In this article we will discuss the following: How can media-based learning and cyber culture create new educational opportunities? How to use this space to achieve successful learning? How can we integrate media and cyberspace into real and traditional classrooms and make learning in these classes attractive? (Pourghaznein, Sabeghi, & Shariatinejad, 2015).

© Springer International Publishing AG, part of Springer Nature 2019
S. H. Sadeghi, *Pathology of Learning in Cyber Space*, Studies in Systems, Decision and Control 156, https://doi.org/10.1007/978-3-319-91449-7_4

Training Goals

Since learning is done through the learner to facilitate and guide the learning of learners, the purpose of learning is used instead of the pedagogical purpose. For learning, a successful education is that it leads to a desirable learning; in other words, it will lead to changes in knowledge, behavior, actions, beliefs, attitudes and values of a learner (Fig. 4.1).

Educational goals, often related to the community and its needs, either address the needs of learners or be a combination of both. Or, in other words, the aims of scientific, political, economic, cultural, etc. are considered. Regardless of the objectives of the training, it should be done in order to provide a desirable training course:

1. The needs assessment is done to determine what the group needs and the purpose is presented accordingly.
2. Present and formulate methods for reaching the goal.
3. Determining the content and tools for transferring that content to the learner.
4. Provide amplifiers for reaching the goal.

To achieve the goals in education, we need to deliver them through coherent and transparent content. Content is something that is supposed to be taught, or what we want to learn. Content is in fact a set of concepts, skills and trends that are selected and organized by planners. At the same time, the content also includes the effects of

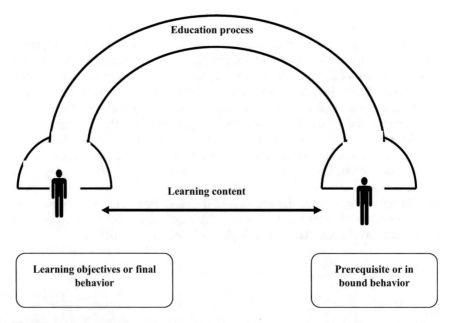

Fig. 4.1 Education process

learning activities. Educational content is presented in several forms, some of which are: text, photos, sounds, videos, charts, information, info graphics, virtual realities, etc.

This content in any formal, informal, and informal education has a special significance and a special place that transmits the goals to the learner with an educational design. Educational design can be defined as specific roles in how to achieve educational goals. In other words, when training is designed to achieve a series of knowledge and skills as a training goal of a set of activities and methods of training, before the training is realized, training is actually done. Therefore, the design of education can be used to prescribe or predict desirable teaching methods for achieving the desired changes in the knowledge, skills, attitudes, thoughts and emotions of learners (Websters Dictionary, 1996), as well as the development of learner capacities for occupation and profession In the future. These types of goals are divided into three categories, which are: essential for learning, useful for learning, interesting to learn. The first step in creating the possibility of realizing educational goals is to select the appropriate and desirable educational content.

There are various goals in reaching the education especially for children but some of them are more important as the following (Chin & Jacobsson, 2016):

- Well speaking, writing, reading and math
- lifelong learners
- have integrity and self-respect
- be able to problem-solve and think critically
- persevere
- be able to look at things differently
- be ready to take risks
- creativity
- community care and willing
- use the world around well.

Objectives of Training in Cyberspace

The formation of cyberspace has affected all aspects of human life, from transportation and communication to entertainment and education. Many educational centers focusing on technological developments now have computers and network facilities. Many of these technology appendices have somewhat embraced the training environment, including the development of computer-based learning and virtual education in the light of the development of computers and, more generally, of information technology. According to recent researches, the Internet has the first place among the twentieth century's inventions and has made space and time constraints in the education debate eroded and at the same time the Internet has made people feel the concept of lifelong learning, Therefore, not only educational centers, but also no one can deny this and reject it (Anderson, Garrison, & Archer,

2004). Despite the fact that cyberspace has spread to all societies, and countries have accepted this fact by computerizing communications but we have not yet fully seen the effects of this media in various areas, especially in the field of education At a stage when John Silicon Brown (2000) called it the gradual development phase of this structured media, we have not yet experienced its explosive effect. We are only experiencing the elementary mode of e-learning, and we have to learn a lot about its core competencies and the creation of a new "learning ecology". What are the relative advantages of e-learning and are they so powerful that they can form a re-conceptualization of learning and learning interactions? Perhaps this is because, because of our inability to recognize the ecology of new learning, no change in education through communication technology has been created, while this technology has made changes in many aspects of society (Anderson et al., 2004). However, the technological advances referred to have made the instructors not only rethink their learning approaches, but also reflect on the need to achieve new learning goals and how to access them (Anderson et al., 2004). In this regard, educational content has also been transformed into the content of e-learning and learning, which in general is defined as:

The e-mail content is referred to as the document set that builds the learner-learner interaction, so that it can be delivered to an electronic format. E-learning content of education is mainly composed of text, image, sound, map, video images, animation, simulation, virtual reality, etc.

There are several general goals in cyberspace learning that can be mentioned as the following (Barnes & Pressey, 2014):

1. Access students at anytime and anywhere
2. Low cost than real universities
3. Possibility of making lessons by professors at any time and place
4. Possibility to attend online classes
5. Possibility to communicate with the professor and other students
6. Ability to use unlimited number of classes.

In sum, despite the changes in education and learning, and the formation of e-learning and education, the goals and objectives of education have not changed. Educational resources in the formats outlined last season have the same learning objectives in real space, which are now being followed by new forms in cyberspace. Of course, it should be noted that education is carried out with different goals and it is not true that it takes into account only specific goals, such as economic ones, and other aspects to be ignored. In education, goals are provided continuously and sometimes at the same time in educational content. But the general categorization we have done for this purpose in this field is based on this view that, despite there are many goals in organizing a course, a class, a story, a photo, etc., ... (something should be added here) One aspect of the goals is more obvious than the other aspects and forms the whole body of the educational content. Based on this, we have set educational goals in accordance with the same general aspect of the content, and it should be noted that the failure to express these goals together does not indicate that they are not included in the contents of the training. Also, here the goals are set free from the educational resources outlined in Chap. 3. This is because each of these resources can have goals in different ways in its own right, and they are used to teach the users and learners, and their content is based on the goals set out in this chapter. In this regard, in order to clarify the educational objectives in the cyberspace, the most important ones are discussed below.

Scientific Objectives of Education

The most important aspect of any training, both in real space and in the cyberspace, is the transfer of academic goals and experiences to the learner. In this regard, with the high capacity and facilities that the cyberspace has, schools, universities, organizations, groups, and even individuals have acquired the ability to diversify their educational content with scientific goals and through existing educational resources. Provide users with cyber space. These goals are transmitted through formal education such as schools and virtual universities, informal education such as short-term training of organizations and courses provided by the private sector and hidden education such as images and games in the cyberspace. They lead to addition of something to the users' knowledge and experience. Although there may be other goals behind these scientific goals, the obvious aspect of this kind of training is the prominence of scientific goals. Among the educational activities that are provided with scientific and experimental goals in cyberspace, it is possible to hold virtual classes of schools and universities using various web resources, video conferencing, e-mail, etc., as well as providing facilities. And the tools of psychologists, surgeons and other individuals and scientific groups in this space.

Political Purposes Training

With the formation of cyberspace, there were many opportunities available to governments, groups and even individuals to pursue political goals through cyberspace education resources. In this regard, each of the groups mentioned above is trying to create and develop the educational resources available in this space for the achievement and induction of their users on the one hand, and on the other hand to use these capacities to reduce the challenges and consolidate their position in Cyber space. For example, some of the political goals in cyberspace can be summarized as follows:

1. Political development of the country
2. Increasing the level of participation of people in political affairs as one of the main manifestations of development

3. Providing the necessary training and expertise for political affairs and educating politicians
4. Socializing the community
5. The possibility of the re-emergence of direct democracy on a large scale.

Electronic democracy can be seen from the capacity of cyber-space to enhance the level and quality of public participation in the government. New technology can make extensive use of new forms of voting, therefore, facilitates direct participation based on elective voting systems (from theoretical perspective) but it has been ineffective (to this day). For example, people can vote from their homes instead of going to the physical place of voting. With the adoption of more polluting and less costly voting systems, people are expected to vote sooner on more issues. All surveys and censuses can be further enhanced, although they need to support a political and organizational will that may not exist in some countries (Hills, 2005).

6. Political objectives of the groups against government.
7. The political goals of the enemies towards the government and the people.
8. Individual political goals toward government, people and groups.
9. Their political parties and goals.
10. Terrorist groups and their political objectives in cyberspace.

Also, one of the greatest political goals of governments at the community level has been the interest of politicians and statesmen in accessing social justice through a more balanced distribution of income. Regarding the effect of education on people's income, the development of educational opportunities as an effective policy is available to governments in order to fit the distribution of income. By publicizing access to pedagogical-cultural opportunities, these expectations have been emphasized that the income situation of low-income families is better and the distribution of income in the whole society is more balanced. Although a single evidence in this regard is not reachable publically, but in some countries, the development of pedagogical-cultural opportunities has been balanced distribution of income which is the one of the important public educational interests. In this regard, since cyberspace places this capacity and opportunity at the disposal of governments to achieve this political goal, this type of training in the cyberspace has a political background.

Economic Objectives of Education

Until the 1960s, the category of education was scattered in two directions by economists: one that educates the quality and value of labor, and other one, the recommendations and economists proposals provided for the educational system. But after the developments and the spread of human knowledge, the economists' attention to education grew rapidly. To the extent that Adam Smith conceived the cost of education as a cost of capital. According to him, the society with educational investments can fertilize the potential of the people and thus provide a way to use more than natural wealth. In his view, trained and skilled labor force will increase productivity, because they are more productive. In line with these intellectual developments, today there is no longer any thought in the belief that education can be considered as an extremely important investment for each individual and society, and further efforts in this regard would be to benefit more from income for individuals and for growth and resulting development then will be for communities. During these changes in the late 1990s, the global economy faced two structural changes stemming from the globalization and revolution of information and communication technology that some analysts came up with. The new economy which came from these two phenomena, became the source of the next changes in the economy. In this regard, the development of information and communication technology and the formation of cyber space as one of the dimensions of modern economics could directly and indirectly impact on the economy, so that it is currently one of the most important goals pursued in the cyber space. Goals are economic. These goals, with the help of the capabilities of the cyber space, signify the conduct of trade, the provision of services, the promotion of goods, the training of the use of goods, the provision of financial services and the exchange of data, etc. Also, cyberspace has important advantages in conventional commerce in physical markets and the way to do business. Today, companies, governments and businessmen are rapidly developing networks in cyberspace to join the cyber-space, and to bring their products, services and goods faster, on time, and at lower cost to their customers all over the world. Accordingly, websites and other training resources available in the cyberspace have become the gateway to a brand, service products, even if it does not have a cyber-buying online company.

Accordingly, given the high capacity of the cyberspace for the economy and organizations, groups and related individuals, a relatively large part of the educational objectives in the cyberspace are related to the economic objectives, which in various ways it is taught to users. For example, training on how to use tools and equipment, how to buy equipment, teaching new technologies in order to attract customers, holding various training courses at a fee, etc.

The Goals of Education at Various Levels of Society

Objectives of Education at the Community Level

One of the main goals and tasks of teaching from the very beginning was social duties and efforts to meet the needs of society. This training, even when the parents of the child and adolescent and the governmental as well as private institutions were responsible, is carried out to achieve this main goal. Nowadays we can see that in many developing countries there is pedagogical and community-based programs and practices played an important role in reaching objectives and strategic plans.

Over the next several years, they will continue to contribute to the improvement of health and safety outcomes in the developed countries. With the formation of cyberspace and the creation of educational facilities in this space, the ground for education and learning in the whole society is provided so that these capacities can be used for teaching purposes in the whole society and a platform for learning and achieving these goals throughout the community is provided. The objectives of education at this level can be derived from the above objectives (economic, political, cultural, etc.), which can be provided through educational resources available in cyberspace. Some of the benefits of cyberspace education through the entire community, with various goals, are:

- Improving economic growth at the community level
- The distribution of income at the community level is more appropriate
- Development of activities and job opportunities
- Inhibition of population growth
- Development and Political Participation
- Inhibition of crimes and social offenses
- Promoting community health and health
- Developing cultural beliefs and attitudes across the entire society
- Creating a positive attitude towards managers and members of the government at community level.

   And ...

Of course, it should be noted that the benefits raised above include a positive aspect of education for the entire community with different goals, and the negative aspects of this type of education can be seen with different and negative outcomes to the benefits mentioned above.

Objectives of Education at the Level of Part of Society

Human society has different differences and perspectives, which can be categorized into smaller sections using various indicators. For example, at the level of a country, people can be separated from each other in the following sub-sections:

   i. The level of economic life
  ii. Scientific level
 iii. Religious groups
  iv. Age groups
   v. Thinking groups
  vi. Groups with common political thoughts.

   Based on this, with a glance at the educational resources of cyberspace, we realize that in this space, according to the needs and views of different sectors of society, specific educational goals are presented that are different from the goals of the whole society and can sometimes be in such a way that induce the goals that in conflict with the whole society and, in some cases, put these groups in front of the community and create challenges in different fields between the levels of society and the whole society. It should be noted that the difference between society and some of its sub-sectors is normal, and the needs, interests and beliefs of these groups are naturally different from other groups of society. Therefore, based on these interests and beliefs, from different educational resources, in order to achieve the desired interests among these groups, appropriate educational objectives are designed and presented to them, which is different from the other groups and the whole society.

Training Objectives at the Individual Level

Although the needs assessment is done to determine educational goals based on resources such as the learner and the community, in many cases the learning needs are considered to be the same as community's needs, community's needs will be a learning need and the learner's need can be studied in two fundamental aspects:

A. Essential needs, which include physical (physiological) needs and psychological needs

These needs are important throughout human life. But often plays a more important role in childhood and adolescence. Suitable satisfaction of basic needs can have a significant effect on mental health, personality development, and life success (including the educational achievement of children and adolescents). Just as neglect and disorientation of these needs can cause many physical and psychological disorders and prevent a person from achieving proper success.

B. Personality development needs

Another category of learner needs can be related to personality development. According to a number of experts, the dimensions of human personality are: physical, emotional, rational or cognitive, social, moral, mental and spiritual proportions.

Personal benefits of training:

- Increase in income
- Increasing the employment of educated people
- Improvement of socio-political situation
- Interests.

Overview

Texts, images, videos, animations, and, in general, any kind of electronic content used in the cyberspace used by website administrators, blogs, portals, etc., is the medium of education and the transfer of concepts and meanings to the minds of Internet users.

These trainings, which can be presented in an official, informal or hidden form to users (learners), seek to realize one or more of the scientific, political, economic, or social-cultural goals.

These educational targets can be designed for a wide range of people. Some tutorials in cyberspace are targeted at the entire community. Issues such as inhibition of population growth, increased political and social participation of the people, promotion of community health and the like are among these. Along with the training that is for all sections of the community, cyber training is provided which mainly addresses the needs, values and views of the section or specific sections of society (such as specific age groups, religious groups, political parties, etc.). In sum, for the design of educational goals at the individual level and its development to the social level, two major categories of human needs, namely physical (material) needs and psychological needs (spiritual), should be considered.

Chapter 5
Pathology Caused by Cyber Education

Introduction

First, before entering into the controversy of government at various levels, it is necessary to state the general explanation of the term "government" and how it is used. The vertical dimension of government is the political structure of the legal organization, which is based on the foundation of public demand and public satisfaction in a specific human group, and the land of its horizontal face is counted.

After a while, the government and industry officials noticed the problem of security of cyberspace and strengthened cyber security. The extraordinary growth of technology has created new challenges that change rapidly with the growth of technology. The progress made in the areas such as computing power, cloud computing, mobile, artificial intelligence, ubiquitous interconnectivity, and Large-scale automation systems is constantly changing, which has raised concern about cybersecurity, personal identity and privacy, and other safety issues. Security threats and compromised privacy can have adverse consequences, and these consequences will spread to the coming decades. Experts have defined the concept of Cyber pathology as follows "People spend a lot of their time working with computers, computer games and the Internet, and spend unreasonably time in front of monitors", This may have physical and psychological consequences for the individual including Sleep constraint, limitation of person-to-person communication, affecting physical activity in the arbitrary space.

General Definitions of Concepts

The government has the following elements or pillars:

(a) Executive or government.

© Springer International Publishing AG, part of Springer Nature 2019
65
S. H. Sadeghi, *Pathology of Learning in Cyber Space*, Studies in Systems,
Decision and Control 156, https://doi.org/10.1007/978-3-319-91449-7_5

(b) Legislature or Parliament.
(c) The judiciary or arbitration between peoples to reach the justice of the community.

The country's political presence is due to the combination of these three pillars and the support of the country's defense forces. Accordingly, coordination between the main forces of the government and the force of the commander-in-chief of the entire Defense Forces is the responsibility of the head of state/government.

But the government refers to the political management and the board of directors of the country and the cabinet of ministers.

Also, the other word that needs to be defined here is pathology. This term is mainly from the field of medical studies to the field of sociology studies and examines issues that damage an organ or a coherent social set. Of course, pathology, especially in the field of learning and psychology sciences, does not lead to the discovery of the areas of injury and intends to consider, along with the study of the aforementioned areas, the ways of treatment and complements to eliminate deficiencies and shortcomings, and finally to increase the potentials for non-slip in future injuries also should be assessed.

Injuries at the Level of Government

Injuries at the International Level

The expansion of the World Wide Web in the twenty-first century, as the most important achievement of Information and Communication Technology (ICT), has eroded many of the boundaries of communication between individuals and governments, and has had profound effects on their performance and relationships. Because of its relatively inexpensive and widely available availability of information and communication technology, it has provided considerable capacity, even for the poorest governments and actors in the region and the world, which may be challenging and damaging. Others should be used. This is just the opposite of the important milestones of the industrial age. In the information age, hardware and software are widely available and easy to use whereas industrial-era weapons such as nuclear weapons, continental missiles, aircraft carriers, warships and tanks are not the same. Therefore, in the information age, governments are not the only international actors who may develop technical capabilities to use for harm. Multinational corporations, nongovernmental organizations, criminal and terrorist groups, and even individuals may engage in war action.

It is true that the information revolution through computers and the Internet has had a great contribution to human progress, no one can in fact deny this fact. Since the transfer of information across borders has been possible, human beings, raw materials, cultural exchanges and diplomatic exchanges are actively taking place across borders, and these international partnerships are covered by a new paradigm called globalization. However, as with the exchange and interaction of borderless information, it is easy to jump over border controls to commit internet crime. Unlike developed countries, which are the largest beneficiaries of the information revolution, the legal system of developing countries that are relatively less favored are lagging behind in terms of technical development. So, they are in a situation where they can hardly face new crimes (international crimes). Consequently, Internet crime that has caused astronomical damage has taken place around the world, but humanity has not managed to effectively control it. However, humanity is experiencing the development of innovative information and technology, and the extent of the Internet crime that exploits this process is increasing day by day (Polański, 2017).

Among these developments in the international arena could be the change in the relations between governments in the regional, international and global arena, which was previously carried out by diplomats. These developments in the field of information and communication technology have transformed this type of relationship between countries and governments into multilateral relations, in which the greatest efforts of the diplomacy and foreign policy system are to establish relations between nations, both through the media and in the cyber space. So that media workers and non-state actors penetrate disclosure and engage in issues by motivating public opinion. These trends have made it very difficult for ambassadors to negotiate with the authorities, because they must engage with many and, at the same time, diverse players. Find a variety of information. It is also in the context of the transformation of many actors into more limited, but more aggressive players, such as non-governmental organizations, which have been transformed by diplomats in the whole world with the ability to adopt organized and collective methods. Also, this technological-scientific revolution has greatly affected world politics, especially the politics of industrialized countries. The above changes greatly diminished the importance of distance and space as the determinants. In line with these developments and the formation of a cyberspace, a new field has been opened up by governments to exploit it at the international level, and opportunities have provided them with harm. A lot of things have been written about government opportunities and governments, but less attention has been paid to the damage caused by cyberspace to governments. Accordingly, we try to make it possible to summarize these injuries in the following:

- Establishing a digital divide between governments

The digital divide refers to the inequality available to ICTs, namely the lack of comprehensive access for the general public to computer networks and other ICTs such as phones, mobile phones, and so on. This form generally occurs when a part of the community does not have the ability to access information technology due to financial and economic issues or the lack of necessary skills. At a higher level, there are some countries, which are not counted at all, have the ability and skills to access the technology. There is no information. This gap, which is currently being developed between countries and their governments with other countries, is growing increasingly between non-developed countries and developed countries, access to ICT and the use of these technologies to improve the productivity and efficiency of processes. Systems and activities in all segments of life, as well as their submission in creating the appropriate infrastructure for active participation in the production of knowledge and information and communication technology and the consumption of tools and digital services, is expanding. Countries can be categorized as follows:

Leading countries; Hotspot countries; Countries that have taken action; Countries that have just begun; Countries that have not yet gone a long way.

The emergence of this digital gap, especially with the formation of cyber-space, has caused injuries such as backwardness, cultural invasion, diminished relationships due to lack of infrastructure for building relationships. Left balance. It should be noted, however, that these injuries can be inflicted by a government, group, and even a person into the desired state and government, given the capabilities and capacities of the cyber space.

- Transferring information to overseas

Due to the development of information technology in the whole world and the interconnection of the global village, spyware is available through the Internet for foreign countries the most important crime that emerged as technology progresses is spyware in cyberspace (computer spyware). Computer spyware can be considered a type of spyware that uses computer systems, and therefore has distinct differences with classical spyware. While classical espionage (whether industrial, political, or military) takes into account the type of information protected and emphasizes the targeting of crime threatened by espionage, computer-based surveillance of the act of custody is contemplated. The advancement of technology has made the cyber-space and its capabilities to maintain the information space of the material space. The prevalence of using this space in order to keep the information turned it into a suitable place for spies to gather information with it. The attention of the spies to this new space as the target of the crime of scourging attention of lawmakers has also attracted this issue in support of this issue and computer literacy has come to the forefront of criminal law (Mack, 2017). This has caused countries and governments try to use different methods, such as Internet spying, hacking governments' websites, to influence the opposing country, and to be able to press the opposing sides in order to attain their own goals. They may also transmit and abuse the information, statistics, and secret documents and security of these countries, whether through individuals or spies in the country of their destination or from other places abroad.

- Driving against government

Since cyberspace is an open space and does not have any kind of single management, anyone can create a specific situation for any purpose. In another word, this space has greatly expanded capabilities to attract governments. Use of cyber-space gives them an opportunity to take advantage of these capabilities in order to wreak havoc on the face of the realities of other countries and other states at the international level and to undermine their power and diplomacy. They can also challenge the country in creating relationships with other countries.

- Rumors and psychological warfare

One of the most important uses of cyber space to harm other governments is creating rumor and launching an international war on governments, so that, by using these rumors and psychological warfare, they can put pressure on the government to pursue their goals on the international level, or to take score from that country.

- Creating inequality of information in the international arena

The issue of information inequality is an urgent concern and as a necessary moderator for many claims about the democratic potential of cyberspace, which in turn presents the themes of cyberspace as the last portrayal of Western cultural imperialism (and especially the American). The language of cyberspace clearly shows this, as Mark Pasteur writes, "the dominant use of English on the Internet, as well as the fact that email addresses in the United States alone do not need a country code, reflects the expansion of US power." The Internet brings American users to the norm. This process of inequality of information mentioned above did not allow the underdeveloped countries to maneuver in this new field and create the background of cultural invasion and other injuries to these countries and governments.

- Among the threats facing governments and governments at the international level, cyber-space education can be mentioned
- Propagating the level of government at the international level
- The direct transfer of thoughts and views of the processes of the system and government into the interior
- Creating a bad face for government in the international arena
- Subrogation and condemnation of the work of government at the international level
- Succession to state and government.

National-Level Injuries

- Reducing Participation in the Social and Political Fate of the Country

People's participation in their socio-political and cultural destiny reflects the importance and trust that individuals have on their government and political system, and the lack of it is an important obstacle to the achievement of development goals. In fact, if the motivation is to become creative, it is necessary for all people to have direct control and direct participation in the destiny of the country and in all stages of the influential decisions. Accordingly, given the possibilities that have been given by cyberspace to individuals, groups, governments to influence public opinion and bring them along with them and also to reduce people's trust in governments. Different ways of hitting a country by making propaganda against it in different ways have reduced the participation of individuals and their role at national level, and created grounds for distrust of government. This reduces the power of governments at national level and the international level, and creates multiple challenges and disadvantages to governments at different levels:

- To create a difference between people and government

The cyberspace allows users at different levels to provide a variety of backgrounds for creating disparities between people and government at the national level. Groups, individuals, and governments provide propaganda, manipulation, rumor and slander, and so on in cyberspace, to create disparities between people and government, and cause injury and challenges for the country.

- Distribution of insulting materials and insults against the heads of state (secrecy transmission)

Unlike a library, information on the Internet is unlisted, edited and uncensored (Gray, 2004). In the cyberspace, there is the possibility of publishing abusive content and incriminating materials to the presidents of the country, which are most commonly found on sites of competing countries, opposition groups, etc. Sometimes it is a special insult, defamation or other misconduct. This is done by opponents with many goals, such as creating doubts among people about the heads of the country, the diminution of officials in the eyes of people in the community, and so on.

- Electronic Publishing Prohibited Content

From another angle, there might be publication of computer magazines with various content, including bombing and terrorist acts, assault and violence through computer and Internet networks. The roller and mindset of the people are marked by the inconvenience of feeling insecure and disturbing the memory. Such activities ultimately shake the cultural infrastructure of societies by altering their insights and shifting paradigms, and then disintegrating government from within. Accordingly, individuals, groups, and governments publish and disseminate these materials in cyberspace trying to create insecurity at the community level, incapacitating government in security, trying to create a challenge and harm national security of the country.

- Creating differences and challenges between local ethnic groups and the government

Ethnic action has been shaped by the multifaceted deprivations of ethnic groups resulting from the distribution of wealth, power, dignity, and the identity axis. With regard to existing platforms that have been created through the cyberspace among people in a country, these moves are intensified through the following three factors:

1. Ethnic elite
2. Interventions of the Aliens
3. Other resources in the cyber space.

In this regard, individuals, groups, etc., pressurize the government in various forms of fire, spark such disagreements over the country, and the drum of inconsistency and lack of justice in the country so that they can excite ethnic and religious groups and turn them into a means to achieve their goals. This in many ways causes various damage and challenges at the national level.

Harming governments at national level through training in cyberspace can be generalized to the followings:

- Creating a gap in the community
- Anti-government propaganda
- Rumor and launching a national psychological warfare
- Leak publishing
- Establishing relationships with subversive groups and subversion
- Possession of parties and political groups who oppose government
- Creating and expanding new social demands from governments
- Disruption in the process of controlling the flow of information
- Creation of pressure groups and virtual penetration
- Spreading deception and managing public opinion
- The excitement of ethnic groups to inequality against state and government
- Dissemination of ethnic and regional demands and increasing the excitability of ethnic groups
- Highlighting local-level differences
- The induction of local challenge
- Establishing unlawful gatherings
- Creating virtual political groups.

Injuries Among Elements of Government

As mentioned above, the elements that make up the government are:

Three executive, legislative and judicial branches.

And the constituent elements of the government are:

The president as head of state, the cabinet and in some countries the prime minister.

Of course, in this section, it should be noted that although the state itself is a member of the constituent elements of the government, this section is considered as a separate section. The damage caused by training in cyberspace on government among the elements of government is as follows:

- Disclosure of private and confidential secrets of government
- Disclosure of private secrets and secrets of government elements
- Creating a challenge between elements of government
- Increasing differences between government elements
- Creating a challenge between elements of government and government
- To create a challenge between elements of government and government elements.

Injuries at the Community Level

Social Injuries

- The challenge of the socialization process

Socialization is a process that is learned and not genetically inherited. A process in which children, or other new members of society, learn the way of life in their community, is called socialization or learning culture. Socialization is the main channel for the transfer of culture over time between generations. Socializing leads to connecting different generations to each other. Socialization is a lifelong process in which human behavior is continuously shaped by social interaction. Socializing allows individuals to cultivate their own talents, and learn and correct (Giddens, 2009). Although knowledge is made in the community, but in educational situations, it is the learner who needs to understand its meanings or to gain a profound knowledge of it. The purposeful process of facilitating access to a socially and individually valuable goal is one of the main elements of the interaction of education and learning (Anderson, Garrison, & Archer, 2004). It achieves it through the presence of the individual in the community and the establishment of various relationships with others and then acquiring the necessary skills. Of course, it should be noted that formal education in the form of a school and university also has a significant role in the socialization of the people. In this regard:

> The main strength of the university is not the transfer of knowledge, but the growth and knowledge that is provided by the network complex and powerful societies.

With the expansion of cyber space and the presence of people on the one hand, and the transfer of training to various purposes, in its formal, informal and hidden forms, the field for communication and relations between individuals and society has been reduced to some extent. More people try to get their facilities and training in the cyberspace. This has led to the formation of a kind of breakdown between the individual and the community. Since most users of this space are teenagers and young people who are less aware of the ways in which they live in society and culture, their isolation from society due to continuous presence in cyberspace has created numerous challenges and disadvantages in the field of "the transfer of lifestyle and social culture" 'for these people'. On the other hand, formal education centers, which have had a real and effective role in socializing, have lost their place in formal education in cyberspace.

- Increasing gaps between generations

One of the undesirable effects, that the use of education in various forms (formal, informal and hidden) in cyberspace can have, is to increase the gap between generations. A generation break is a profound gap that results from interruptions and intergenerational relationships. In other words, when the sources of identity are not mutually reinforcing generations and do not seek a continuity of the generations, the ground for the emergence of conflict and generation breaks is provided, so that members of the community are in a state that can observe their identity is in negation and opposition to their ancestors. This does not put us in the context of cultural transformations—whose existence is vital to the dynamism of society in various fields—but leads us to the margin. A margin that is very costly and controversial that provides the ground for social and cultural collapse. Because modern technologies are more appealing to young people, they will be more likely to learn and develop in this field, and older people will not have much in common with this new technology and will not be in cyberspace. This will make the gap between generations more advanced with the next generations, eliminating the generational gap, and let our younger generation go away from national and Islamic values and its predecessor. Also, due to the younger domination of the use and productivity of the cyberspace, the crisis of control is being created by older people, as the control of students, learners and users in cyberspace became harder by parents, teachers and … This generation gap between people in different ways creates some challenges for society and causes damages to it.

- Loose the foundation of the family

The family is the foundation of the community and the core of maintaining the traditions, norms and social values. The family-based foundation of the social ties of kinship is the focal point for the emergence of human and spatial emotions for social upbringing. Social values are one of the most basic elements of a social system that can be controlled by society and lead to decline or excellence. Therefore, families and authorities should be very keen to know about values. What factors bring them in a society and how they can be reformed altered. On the other hand, the family provides a wide range of emotional, physical, and caring services to its members. Weakness and inability of the family are the source of many problems that social welfare faces. Parents in formation of the personality structure of the adolescent and his knowledge, attitude and practice play a very important role. Accordingly, since the family is the first and, at the same time, the most important source of the community, we describe the Internet-related injuries and training in cyberspace in more detail.

Entertainment in the cyberspace and prolonging the use of the Internet are associated with the isolation of the user. One needs to minimize other relationships in his everyday life to be able to spend more time surfing the cyber space. The Internet, with its felicitous and virtual conditions, especially for young people, can disrupt all aspects of their lives and move them away from the family and go deep into isolation and loneliness. The internet deception in the family environment has led to the breakdown of marriage, marital relations and relations between parents and children. The Internet has diminished the role of dialogue and thought among the members of the family and has made a serious alarm for human communication. The high use of the Internet is linked to a poor social link. Unfortunately, the other cyberbullying education on families, the formation of family conflicts and fights, and lack of emotion among family members that shake the pillars of the family and compromise the mental health of its members. Family traumas and conflicts are a complex and multifaceted phenomenon and should be studied from a variety of psychological, social, economic, legal and communication perspectives. Accordingly, today, each family member goes to his own room and entertains himself with a computer and the Internet. Even drinking a cup of tea or having dinner and breakfast are done on a computer desk. The decline or lack of family conversation will surely reduce the number of family members from one another after a while, and it will lead to a reduction in the level of trust among family members, which will result in negative results. On the other hand, this activity of family members, after a while, leads to detention (especially adolescents) and reduces their interest in social life and family engagement (Fig. 5.1).

- Education time wasting

While Social Medias have helped students to contact their teachers easily and share scientific more than anytime, but according to pathology of social networks they can take much of students study time. Spending lots of time on social Medias and specially messengers like WhatsApp, Telegram, etc. can rises postponement on their studying, increases spellings and grammatical issues on sentences and deficiency of attentiveness during lectures. Continuous online activities on this spaces cause students cannot finish their assignments following their private studies schedules (Yeboah & Ewur, 2014).

Other cybercrime injuries include:

- Easy access to prohibited resources and indifference to values and norms in individuals
- Reducing human relationships and the educational effect of formal education resources in cyberspace
- Reducing social spirits
- The development of feminist values and opinions, and so on
- Insulting different ethnicities.

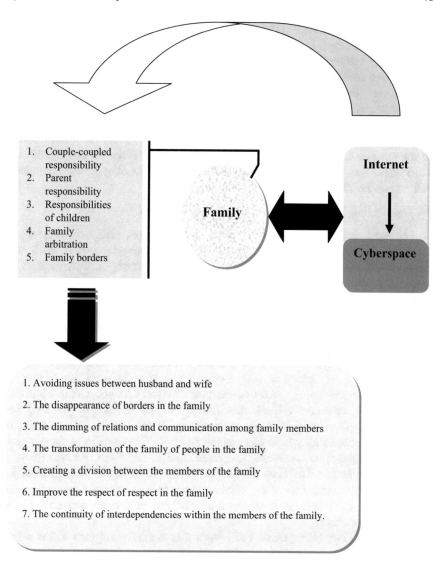

1. Couple-coupled responsibility
2. Parent responsibility
3. Responsibilities of children
4. Family arbitration
5. Family borders

Family

Internet

Cyberspace

1. Avoiding issues between husband and wife

2. The disappearance of borders in the family

3. The dimming of relations and communication among family members

4. The transformation of the family of people in the family

5. Creating a division between the members of the family

6. Improve the respect of respect in the family

7. The continuity of interdependencies within the members of the family.

Fig. 5.1 Cyberspace and internet damage to the family

Cultural Injuries

There are several definitions for culture. Culture is a collection that can be thought of as everything, including knowledge, beliefs, arts, ethics, rights, customs, and other abilities that a member of the community must learn through learning. In general, culture is a method of public life and is learned and mimicked through social interaction and transmitted between generations. Culture of every society has a significant value and a significant position and forms the basis of the formation and harmony of each community. In this regard, societies have been working in different ways to preserve and transfer their culture to later generations, as well as to flourish and grow it. With the development of information and communication technologies, the context of the transformation and challenge of cultures in the community, especially culture in the country (national culture) has been provided. In the meantime, there are many facilities and capacities the cyber space in different formats and for different purposes has: the ability to change and damage the culture of the country significantly and in this way the society faces various injuries and challenges. The cyber-related injuries and education in this space on community culture can be summarized as follows:

- Transferring Western culture and undesirable values

The extensive presence of information and communication technologies that cross the national boundaries of the world, and undoubtedly impose their own thinking and lifestyle on their users, can impose many changes. This leads to the fact that people's cultural inclining is not only in the physical and tangible surroundings around them (through the family, schools, community, and so on), but also in parallel, they also benefit from virtual culture and adhere to it. In this regard, cultural elements are introduced with the help of their information technologies, and individuals are confronted with a new dimension of culture, and cyber cultural pedagogical resources as a mean to transfer the alien culture into the culture of community. The acquisition of alien elements and procedures, doubts in the elements of traditional culture, the re-formation of awareness and cultural transformations at the community level. Accordingly, in the absence of planning and adopting a proper coping with this vulnerability, it can lead to cultural transformation. This vulnerability is so clear to the countries that sometimes some people oppose the Internet and satellites in general and demand the removal of these elements from human life. They may suggest some ways to the correct use of the network and important role of the cultural result to the culture of this space. For example, one of the damages caused by the transfer of culture was pointed out to be the alien to our culture.

Valentine's Day: International Valentine's Day among young people has seen a significant upswing in recent years. Valentine as a cultural symbol of the West has been attracted by young people and adolescents in cyberspace, thanks to the technologies, especially the cyberspace. As a result, many of the cultural values and beliefs of society such as adherence to the Islamic standards have been eroded in some people.

- Pornography prevalence

Another potential damage to cyber-space is the proliferation of pornography that directly creates the conditions and training needed for direct and indirect sexual stimulation in the cyber-space, and has many harmful physical and emotional effects, especially for teens and young people. Punishment is seriously damaging in three respects: first, it is possible for children and young people to access harassment-free pornography on the Internet. Second, bloggers find child pornography and teenagers and young people most likely to be cybercriminals finding the easy way to sell their products-so that sexual abuse can be greatly spread. The third and peril damage to the cyber-space users is that cyber-space criminals can also capture their victims through the Internet, by e-mail, messengers and etc. to finally capture them in the real world. In the meantime, the cyber space is the most important source of propaganda of pornography due to three characteristics. These three features are:

1. Easy access for users to it
2. The ability to pay for sites that using them requires payment
3. The anonymity of its consumers
4. Promotion of alien language among youth as a symbol of Western culture.

Language is one of the most important cultural symbols of any country, nation and even anyone. Because in cyberspace learners should interact according to language and communication methods, as well as written and unwritten protocols. On this basis, it should pay attention to how the language, the communication method and these protocols, are created or chosen and become commonplace? The answer is: It defines the framework as the dominant culture and the pioneers in the field of cyber space and cultural and linguistic authorities of governments should know that. Unfortunately, however, this state of domination of the language and culture of the leading countries in the cyberspace has led young people to become more inclined to learn them than to learn the rules and literature of their national language. This has led to the promotion and development of fluent cyberspace among young people and adolescents in the community. It should be noted, however, that these effects of cyberspace on national language are not limited to Third World countries, but also of industrialized countries. Jacques Chirac in 1996 stated, "English language dominance over the Internet and the cyber space has caused the culture and language of France and other major languages of the world to be destroyed".

• Getting away from Islamic values

Inappropriate use of technologies, such as the Internet, without culture and appropriate education at the community level, can lead to the younger generation of the community being discarded from national and Islamic values. The two factors that are known in this regard are the transfer of Western culture, as well as the creation of a gap between generations, as mentioned above.

• Formation of virtual culture

The development of cyber space and its growing trend have shaped the culture of this space in different ways in the users. The development of this virtual culture means leaving the real world, and equally, away from structures and standards, which itself creates the basis of the formation of many harms in society.

- Cultural tensions

Building the cyberspace, the Internet can bring the context of cyberspace to a contraction of ideology and cultural tension in a global context. In a situation where people at the community level have heterogeneous cultural ideas, some cultures, instead of more tolerance, consider themselves superior to others and seek to induce their culture over others (Houston, 2001). This in many ways causes the challenge and damage in the community.

Among the damages caused by the Internet (due to cyberspace) on culture, one can mention the following:

- Change in dressing and making up
- Change in eating habits
- No monitoring of cyber curricular resources
- Promoting corruption and indifference among people through: photographs, films, vulgar and pornographic content, creating illegitimate relationships through dating sites, chat rooms, etc.
- Trying to digest subcultures and eventually destroying them by promoting their desired culture.

Political Injury

- Control and guidance of public opinion

Public opinion refers to just a simple statement as a previous warning of an event. Such as: "Country A has tested a strong bomb". Public opinion as a mere belief is not a prior knowledge of reality, but its evaluation is actually a prediction of the future flow of events. Such as "there will not be a civil war in this country". Emphasizes public opinion as a flow of action from an action, for example, when asked, "Should country 'A' fight for its own interests with country 'B'?" It implies that pursuing an action should be surely considered (8). There are many types of public opinion: the public opinion of the small community, the public opinion of the nation, and the global public opinion (9). The importance of public opinion is determined when people, politicians and religious leaders generally have to assess the influence of public opinion in order to be able to consolidate their position and to achieve their security and prosperity in the country and in their sphere of influence. Based on the pattern of public opinion in the country and society, we can predict behaviors of people. The ability to predict the future behaviors of the people is of great value, because it gives individuals as well as political and religious leaders the opportunity to push their efforts towards their constructive and beneficial objectives. The reality is that human beings live in the same way that they live with their thoughts and opinions, and that in human actions we realize that thought and vision play a more important role in the truth. In this regard, with the high importance of public opinion, it can be said that being able to mobilize and controlling and directing it can give a high power in political arena at various levels (local, national, and international).

Today, the use of the Internet, through employing the capacities and capabilities of the cyber space, has been able to provide a prominent role in controlling and directing the public domain (public opinion). Now the undeniable role of cyberspace in creating and shaping public opinion and influencing existing resources in it with different educational goals on social and political developments is almost universally accepted. Based on this, by using its cyberspace capabilities, it has become a powerful tool for controlling public opinion. It can affect public opinion, and predict their future behavior. That is, they will be aware of what people will need in the future, because perhaps they themselves have inspired these demands that ultimately lead to the homogeneity of societies and the elimination of constructive contradictions, which are the main factors of the dynamics of societies. This leads them to the desired goals, which is a strong support for every political movement at its various levels.

- Modeling

One of the ways of psychological warfare in creating or exacerbating political crises in the community is "imagining" through cyberspace. Guiding by helping the process of long-term media impact especially the Internet plays a major role in the success or failure of actors during the crisis. A collection of existing images in different aspects of reality can defines a structure of new concepts in the individual's mind. The function of imagining in the media is through various educational resources through changing the image of the news, real images, changing events, contradictory content, and so on. To change sights of people. This first change affects the feelings and imagination of the people and then distorts some old ideas in their minds. The entry of damage into the archive of previous images allows for the adoption of new images. This method is a media outlet. Meanwhile, since the cyberspace has a lot of potential and facilities in creating this imagination among the people of the community, it has been able to use its ability to formulate the desired ideas by distorting past perspectives and replacing the notions. Instead, they will achieve their goals. One of these ideas, created by opponents in the cyberspace and affecting the people of the community, is the idea of distrusting the government and the elements of government among the people, as well as creating distrust among the people of the community. The formation of this type of distrust among the society with the government causes the formation of numerous political damages in the society, for example, the lack of importance of the community for political change in the country, lack of participation in public opinion, lack of political support of government in carrying out projects and …

Other cybercrime-induced political injuries include the following:
Made:

- The damages caused by creating religious gaps among people
- Disadvantages caused by ethnic clefts among people
- The damages caused by creating a gap between people and political leaders.

Economic Injuries

- Creating a heavy economic burden for the family and society

The use and productivity of e-learning and cyber space in formal form, for example through universities and schools impose a great deal on the society and family, especially in the third world countries due to their need for hardware and software equipment, multimedia content, extensive training for stakeholders and cost-effective economies and leads to the formation of various harm in these societies and families economically at the same time.

- Advertising of foreign goods and questioning domestic goods

The capacities and capabilities of the cyber space give companies, organizations and individuals the opportunity to expose their products and achievements to the views of people in the cyber space. They can also sell them in this way. In this way, the widespread and continuous advertising of these products attracts users in the cyberspace and earn money.

- Propaganda
- Facilities
- Luxurious.

The cyberspace has the conditions in which they can display and offer luxury goods and luxury items in different ways, and this makes available an opportunity to promote the luxury of culture in different societies among young people, especially in the Third World Societies, transforming it into a culture among individuals, and thus achieving widespread economic benefits. The existence of these attitudes and economic uses of the cyber space causes various economic losses in societies and introduces various harm to the economy of the family and society. Among other damages in this section are the following:

- Damage caused by fraudulent economic education
- Generating high costs of using the internet in the family.

Scientific Injuries

- Losing Learning Opportunities
- A professor has the most influence on the formation of the learning and learning environment and its goals. Objectives, content and evaluation are largely done by the professors. With the volume of information and ease of access to the ocean of information, the main task of the professor is to draw a path through this turmoil, create discipline and conditions for strengthening deep approaches learning was created (Anderson et al., 2004). But with the formation of e-learning and student-centered learning of these systems, the role of the teacher in guiding and arranging the way educational resources is used becomes less than before. This results in losing learning opportunities and students become more limited and less well off, and many creativities that are possible in the relationship between the teacher and the student cannot flourish, which leads to scientifically harmed individuals.

The level of student reciprocity and the habit of copying others the cyberspace has many articles and books and we can easily find in a book or article by searching for the topic in question. This causes students to have no need to study and understand the whole book and article. Also, excessive capacity of the materials they provide, makes it easy for them to copy the work of others and deliver them with their names. Providing the possibility of this kind of abuse, due to the lack of supervision by teachers and professors, causes the students, who use and e-learning-based education to be on a superficial level, to have the habits of mis-using and copying of work of others, instead of scientific production, these habits grow in them enormously.

Of course, despite the sensitivity and importance of this issue, enough research has not been carried out by researchers and experts on the damage caused by e-learning and education in cyberspace, and commenting in this field requires the use of field research. In the following we mention the two well-known, well-proven, injuries.

Injuries at the Individual Level

Psychological Injuries

- Confusion

With the formation of cyberspace and the creation of high-capacity facilities in providing information in a variety of ways and with different goals and fields, individuals face the ocean of information and content that includes recognizing the truth from non-truth, the truth of falsehood, the validity of the lack of credit, the truth of the falsehood of the contents, and ... confuses the individual. This person's confusion can provide the grounds for his harmful use of inaccurate information on the one hand and his frustration on the other, and heighten the background of the formation of mental illness in a person.

- The diminution of social communication, isolation and individualism

One of the other effects that the presence of technologies, especially the Internet, and the formation of a cyberspace in lives of people can have, is the reduction of real social communication between individuals, making self-isolation and individualism. However, education and learning in the real world, especially formally and informally, strengthened the social relationships of individuals, but the formation of cyberspace and virtual world has reduced social relations and active communications. This causes self-isolation and increasing individuality in people, which can cause various forms of mental disorder in a person.

- Violence

One of the major concerns of cyber-training, especially video games in this space, is obviously the violent nature of many of them. Concerned about the effects of games, in line with the growing public concern about the effects of violence in the cyberspace, is due to the fact that these games are interactive activities and one finds itself part of it when it comes to cyber space. Therefore, it is possible to involve the mind and emotions of the person more. Hence, the influence of violent themes of these games and this space on individuals, especially young people, is more than other instruments.

The internet had been known as the important part in 20th century in daily life for several globes. Many usage such as learning, shopping, working, meeting, online communication had been made this technology as an increasing affordable, so there is also a new threat that has brought us to an interesting moment, where the issue of sexualized threats in online public spaces is a concern facing more and more women (Owen, Noble, & Speed, 2017). According to general irritability theory, games in cyberspace increase the level of stimulation of individuals and in practice, they add to the energy and intensity of their emotional action. Thus, very violent games in the cyberspace can lead to aggressive behaviors in the individual and cause psychological damage.

- Harm of Internet addiction

Addiction not only makes victims to destroy themselves, but also makes them Secluded and far from friendships and intimacy. Internet addiction is no exception and causes physical and psychological damage to the individual. Internet addiction is a subset of impulsive disorders, which leads to creation of psychological, social, educational, or occupational problems in a person's life. The most common definition of Internet addiction is to create a kind of behavioral dependence on the Internet. The main criteria for addiction, which should last for at least 20 months, are: tolerance, signs of the syndrome (such as shivering, tremor, anxiety, repetitive thoughts about the Internet, imagining and dreaming of the Internet, voluntary or non-volatile movements and the willfulness of fingers), the sense of compulsion to use the Internet to reduce or prevent the symptoms of withdrawal, to use more than the time it intends to use, to reduce social, occupational and recreational activities and the risk of losing jobs, education and job opportunities because of more use of the Internet. Based on this, Internet addiction will begin with the results for the person. Drug addiction may be one of the main problems of psychological problems in a person, including Internet addiction problems. Caving and isolating individuals in cyberspace will cause a person to lose social and career opportunities and opportunities. This process of addiction to the Internet has serious similarities with drug addiction, and despite the differences between these two addictive factors in terms of "material building" (one is chemical and consumable, and the other is merely a communication), both are responses. They are different from the same requirements that go back to the same pathological background. In this regard, an injured person who has a socio-psychological background to drug abuse, is exposed to drug addiction if exposed to drug access opportunities in peer and not controlled groups.

- Closure and development of social skills

With individual isolation from the community, due to excessive use of the Internet and presence in the cyberspace, the person finds little opportunity to attend the community and various groups to learn and gain social skills. Individual relationships are limited to his relationships with other people in cyberspace and in the family. This makes it impossible to gain the skills necessary to communicate with others, to live in the community, and in real life, they have many problems and challenges that can lead to psychological harm in one person.

Depression, psychological trauma due to the misuse of information in the private computer of the individual are other psychological injuries that can be caused by inappropriate training and learning in cyberspace.

Physical Injuries

Since the presence and use of training and learning and the productivity of the Internet and cyber space require a long time at the foot of the computer, users of this environment have low levels of physical activity and movement. This low physical, physiological mobility of the body, and even in terms of brain processes, causes changes in individuals and provides different backgrounds for physical injuries to the user. Among these physical injuries are the following:

- Eye injuries
- Wrist injuries
- Lumbar injuries
- Other physical injuries
- Obesity in the development of neck diseases

- Development of Foot Diseases
- Gastrointestinal tract damage
- Damage to the skeletal system due to continuous and non-standard sitting in the front of the monitor.

Conduct Injuries

- Courage and daring to commit crime because of the unknown in the environment

Criminologists say one of the reasons for the crime in the margins of the cities is their immigration and the unidentified people of one neighborhood for each other. It has been proven to be that if people know each other's location, they will commit less crime and harm to each other. Meanwhile, since cyberspace is a space that its inhabitants do not know in the same relationship. Every day it gets into this space with a special identity. Therefore, this space itself becomes a criminal and harmful atmosphere, and people are courageous and daring to see that they are unknown, trying to harm others and offend them. In physical world, if the police, after a lot of work, could find the address of the offender, he faces a real area with geographical boundaries and restrictions. If the offender is in another country, it would be hard to arrest and punish him.

- Loosening of religious beliefs

Values and beliefs, and at the top of their beliefs and religious beliefs are among the most important elements of human personality. Religious orders are involved in all aspects of life and are considered as factors in the growth of personality and determine the interpersonal and social relationships of individuals. In this regard, Islam has a special interest in organizing interpersonal and social relations. The existence of multiple sites with the goal of introducing religious beliefs and sect delusions and challenging beliefs the religious person in the cyberspace is losing religious beliefs among users, many of whom are young and adolescents who do not have strong beliefs. These educational sites, by relying on their so-called logical training, try to prove many religious issues with logic, and with obscure examples, the creation of a fallacy and giving a world view to people, the make basis of the formation of doubt among people. They bring up various questions about religious beliefs of individuals and finally, by giving answers to these questions in general and then with some coverages that these answer truly make sense, they establish bases for training their goals in the individual and then create religious beliefs in a person. As a result, the person feels less comfortable under conditions of a well-planned and educational environment.

- Decreasing respect for each other

Among other cyber-space injuries, one can mention the reduction of respect among people, especially those of older adults. Some of the educational resources in cyberspace by presenting specific stories, neglecting values, ignoring the experiences of the elderly, simplifying relationships, etc., The formation of beliefs and the spirit of dictatorship make people less relevant to their relationships with others, especially older people and parents.

- Decreasing ethical principles (trusteeship, commitment to duty and honesty)

With the weakening of religious beliefs among people in the cyberspace, on the one hand, and the lack of respect for values in this space, the context for damaging and neglecting ethical principles among people increases and individuals simply adhere to the rights of others, and as a result, trust and other ethical principles are violated.

In this growing knowledge, instructors always hope that academic dishonesty effects on unethical behaviors such as cheating and plagiarism, which are known to be resistant to extinction. In this regard, despair is not the answer; therefore, teachers can have effectiveness on the occurrence of unethical behaviors among their students. So there should be a conscious and deliberate effort that can lead the create climates that encourage ethical student behavior is possible and can be successful.

Establishing Unclear Relationships Between Individuals

With the provision of uncontrolled, widespread educational environments for various purposes, cyberspace provides users with a variety of relationships that provide them with the excitement of individuals and the creation of a wider range of relationships. These relationships, which are noticed beyond the limits and uncertainties of individuals in the cyberspace, create and develop specific views in individuals. This process of forming without fear and boundary with other people, and especially the opposite sex, creates a wide-ranging and often unpredictable relationship among individuals, whether in the cyberspace or in the real world. However, this can cause multiple physical and psychological injuries to a person at a higher level.

Identity Damage

Identity damage is a part of the damage to the community as well as personal injury, but because this type of injury is very important, we will examine it individually. In Dictionaries identity is defined as "Individuality" which means "possessing a set of characteristics" that are distinct from others, making it possible to identify and differentiate an individual from a person, a group of people, or a citizen from another, and all members. It likens a group and people. In other words, conceptual identity refers to personal states and practices, and has the root of family education, cultural learning, and social beliefs (Burger & Luckman, 1964). Identity consists of two components: cognitive (mental) and psychological (emotional and psychological). The mental part (beliefs) consists of a set of values, beliefs, norms, symbols, and attitudes (attitudes), and the psychological (emotional) part of the set of emotions, affection and hatred, tendency and despair, belonging, commitment, and duty … The presence of these components and awareness of them lead to the formation of "self" in one person or a community (group, society). Therefore, identity at the first level is divided into two levels: individual and social.

A person has his own goals, interests, and special interests in the areas of "interpersonal" relationships, "religious beliefs", "occupational status", "future life" and attitudes towards oneself, which make him a single person. Based on the difference, individual identities are distinguished from other individuals (within the group). But identity at the social level is in the sense of collective identity and group identity, which separates a group from other groups. That is, the set of features that exist in a certain number of people who create "us" and distinguishes this group preferable than others.

Given the prominent position of identity in individuals and society and taking into account the changes that have taken place in the contemporary world, the phenomena of the development of information-communication technologies, and the formation of the cyber space, followed by the increase of users have caused major changes in the values and attitudes about life and created identity changes, especially in the youth. These changes are to such an extent that with the advent and expansion of cyber space, identity, its meaning and basis have changed and from a relatively stable and integrated phenomenon has become variable and black a phenomenon and several pieces.

The same problem has caused damage and, in some cases, even provided the basis for the destruction of individual and social identities in society. Based on these changes, the identity issue is strongly disputed in social theory. Also various discussions around it have been formed. In the meantime, the main argument is that the old identities that kept the social world for a long time are declining and new identities that fragment the modern human mind and personality as a single subject are emerging. This is the so-called "identity crisis", which has been regarded as part of a wider process of change that is shifting the structures and processes of modern societies and shakes the framework that gives people the fixed support in the social world. Therefore, the cyberspace, through its social interaction, is capable of creating conditions that can be effective in organizing the path to the interests of individuals and the formation of new identities. Thus, the social foundations of identity seem to be radically changing in the cyberspace, and new identities are emerging through online interactions, which are called virtual identities. These identities are very fragile and unstable.

The challenges posed by this debris appear in different ways. One of the most important indicators of socio-national identity is the interest of the society in its continuity and its integration and non-separation. Now, if a national identity is in a society with a challenge or a crisis, the legitimacy of that community's life is weakened, and the desire for separation and social segregation among the citizens of such a society increases. With the upcoming trend, which is an important part of the identity of society, especially among young people and adolescents, has faced serious challenges. The ground for the weakening of national identity has also been provided to individuals and society. Today, young people are less familiar with their national identities than in the past, and even in some areas they prefer new virtual identities to their national identities, which themselves have seriously damaged national identities.

On the other hand, one of the occasional requirements of Internet presence is lying and concealing its true identity. Because this space is bad enough; it is unknown and anyone entering a real cyberspace can be abused. So people prefer to hide their real identity. Now, since most of the users of this space are young, these secrets and lies are at odds with the main feature of the young age that identifies. A young man who seeks to discover values and internalize them is antisocially valuable, which can be very dangerous.

The processes mentioned above make it difficult for a person to belong to the community in which he or she lives, and do not properly play their social role, and, as far as possible, diverges from their national culture and values. This seriously affects the identity of the individual, society and national identity, which in many cases, with the continuation of this process, may become an identity crisis in the community. Of course, one should not be ignorant of the role and position of the educational environments in the cyber space, which would provide the user with the desired identities and characters for specific purposes. These environments play an important role in avoiding a person's real identity and raising doubts about it, and in some cases hatred and pessimism towards it.

The Injuries Caused by the Training of Social Networks in Cyberspace

The Internet, through which the formation of the cyberspace provides opportunities for the audience, actually helps the audience to become powerful in controlling communication. With the help of the other Internet audience, it is not passive to just watch the content of media programs, but it has the power to show feedback quickly with content, to influence the way content is produced, and even to produce content themselves. On the one hand, groups and movements that have previously been unable to represent themselves and contribute to the production of content in the media and have been marginalized by existing power relations have, to a certain extent, been able to make their voice and presence listened to this space. On the other hand, virtual social groups, formed through relationships and interactions in this space, also have the opportunity to express themselves and make opportunities for themselves in different fields with different goals.

Technology may have helped you in many aspects of life, but this does not necessarily mean that technology is for the health of your body and soul. It is true that man cannot be away from technology today experts believe that our dependence on technology can be harmful and even dangerous. Even recent studies have shown that technological tools are available to the public that many people cannot use with words without them, they know, they can have harmful effects on our health (Giddens, 2009).

These groups can defend their rights in this space and react to representations and faces shown by others, or to reveal their faces through this space to others. Also, these groups can accelerate the circulation of information by increasing the volume of its information due to the global nature of the cyberspace. In addition, these groups can use cyberspace facilities as educational and advertising tools for the propagation of their cultural, social, or political activities. Accordingly, given the facilities and opportunities provided to these groups, the context of the occurrence of some injuries and challenges at various levels through these groups have been created which some of them are mentioned below:

- The emergence of social movements

As economic relations tend to be more mobile and less voluminous, civic symbols and patterns are also driven by cyber-space through the exchange of information during a continuous change. By transferring our information in a networked society, internal dynamism is constantly shaping new social and political structures. Understanding this condition, Alan Thorne announced that the era of the great revolution has passed, and social changes are now occurring on the bedrock of social movements. The social movement is not a collective phenomenon, such as the protest of a political party; since a party, if weakened, is uncoordinated organization, but a social movement has the ability to renew itself again in the event of a loss of organizational cohesion. These trends are expanding through cyber space. Virtual social movements are the interactions between different actors, which may or may not include official organizations, as the case may be. Therefore, social movements are not members, they are contributors. This engagement emanates from the feeling of being involved in a collective effort without having to formally belong to a particular organization. In this way, the person's participation is based on his individual desire. In DeLaport's view, it is clear that the decisive division of the movement's activists into organizations with monopoly paradigm frameworks causes the activities of the movement to break through and, as a result, the dissociation between activists of the movement takes place. Conversely, the existence of organizations with cultural aims exacerbates interactions within the movement. And as a result, a mass mobilization of campaigners takes place. Many social networks that are created by websites or blogs can be arranged with the information and excitement of members of the assembly or gatherings for specific purposes. For example, students can create student movements through virtual social networks in this field.

- Education of negative tendencies in society

The formation of groups and sects-atheists-with specific goals and motives, which are now Satanist groups and many others among the atheistic groups-are also horrible and superstitious, and so on. It has become prevalent in this way.

- Teach people to infiltrate and steal confidential documents
- Encouraging members to attend unofficial gatherings and demonstrations for specific purposes
- Establishing coalitions to hit the government
- Creating insecurity to achieve its goals and gaining a place in the ruling power
- Developing urban rumors
- Creating sectarian perspectives and propaganda (racism).

Security Injuries

- Training of malicious groups

With the advent of the information and communications revolution, governments have been in various forms of commerce, administration, and ... cyber-space. This ability of information technology that has created cyberspace has not been left to governments. Groups and others have also used these facilities to reach their goals. By using the cyberspace, these groups train individuals and other destructive groups in space against the countries and the things they are seeking to achieve their goals in this area. Of course, this issue is not limited to these groups and individuals, and the enemy states also use these facilities to achieve their goals through training their pro-democracy groups and the opposition of the desired government. In this context, we can discuss the training of hackers who attack the confidential and security information of the concerned countries, and in various aspects of politics and ... used by governments and groups to achieve their goals. In this regard, by developing of technology, many governments have responded to electronic communications crossing their land borders by stopping or streamlining the flow of information. However, because of the unknown, this space and facilities on the one hand, and on the other hand, the growing need of governments to use and benefit from these spaces, the security threats to society and governments posed by the training of destructive groups in this space continue to exist and are expanding. Because of the high cyber-space capabilities, vulnerabilities in the world are completely predictable, and anyone anywhere in the world can only have the ability to detect and have the necessary tools to do so. Attack and damage the cyberspace of the country. Thus, computer attackers are able to rack up national networks without warning, and developed so quickly that many target positions, even the opportunity to hear the siren sound, they do not find a peril, and even in the event of a previous alert, they do not have much opportunity to defend.

- Information War

The war of information is part of the intelligence operation, including the military activity that occurs during the war and is carried out in the information environment. By developing the Internet and the formation of the cyberspace and the role of this space in various information fields, the current information warfare has become one of the most important parts of the cyber warfare in this space relative to security, especially government security. Concerns about foreign attacks and information warfare are so high that even a country like the United States, which itself speaks of information technology in the first place, has been pushing for a solution.

Today, most countries have plans to protect their information systems, and some also plan for intelligence operations. A set of invasive capabilities of information operations is as follows:

1. Attack on computer networks: Operation to corrupt, remove, shake and disassemble the information of computer or computer networks of the enemy.
2. Electronic warfare: Any kind of military activity that uses electromagnetic waves and concentrated energy to control the electromagnetic spectrum.
3. Psychological Operations: Operations that are planned to transmit selective information to people from foreign countries to influence governments, organizations, groups, and even foreign entities.
4. Military deception: A deliberate operation used to mislead enemy decision-makers.
5. Computer Information Warfare: Some lecturers believe it is not possible to attack large-scale virtual space anymore, the surveys show the growth of attacks.

- Drugs

Easy access to the Internet, through which the cyber-space has encouraged many individuals and groups to use this device to promote or facilitate illicit drug-related activities. Training and information that facilitates the production of narcotics in cyberspace include materials on training equipment or other resources needed for its production, drug training facilitates education, including methods of drug use, works, propaganda of new drugs. Also, information that facilitates the sale of drugs and marketing, including how and where drugs can be made, or the mechanism for online drug purchasing.

Of course, these individuals and groups are not necessarily producers, smugglers or drug users (although they may be) and may not explicitly encourage engagement in illicit drug activities, but what is certain is that they are usually trying to exaggerate and idealize the use of drugs.

- Women's exploitation and trafficking

Even though women's exploitation and trafficking have existed since the past, the Internet is one of these activities it has made it easier for law-makers to create illegal educational and friendly environments, and has led cyber-space users, most of whom are young and adolescents, to be trapped in trafficking and in these environments. There are many things to see.

- The dissemination of confidential and forensic documents to create insecurity

Usually in cyberspace, banned content and security is easily e-mailed. The methodology also varies depending on the use of the mailing list, email, and newsgroups. Finally, the forbidden content that has its own security is on the network. This is done by groups, individuals and, in some cases, governments to create insecurity and anxiety at various local, national, and international levels for specific purposes. The most recent and most important feature of this is the WikiLeaks site, which with the specific goals of the owners of the site and behind the scenes, created a background of security and turmoil in the global environment.

Among other training that can provide insecurity through cyber space can be:

- Organized crimes, including economic corruption, to create insecurity in the country
- Training in the construction and sale of terrorist and sabotage tools
- Relationship between destructive and impostor groups
- Creating a suitable platform for the activities of opposition groups
- Stealing electronic money, stealing credit card information, extortion, and deception.

Chapter 6
Reduce and Deal with Injuries by Training in Cyberspace

Introduction

The Internet can be surely the most important and prominent feature of information and communication technology. The World Wide Web as a key component of the Internet has made it possible to publicly provide text, audio and video on the Internet, with high speed data and information exchange, having multimedia environment and ease of access. The existence of an Internet system, its lack of belonging to a particular person, its permanent activity and its lack of shutting down, and the low cost of traditional commerce can be considered as some of the most important reasons for the massive acceptance of this tool for e-commerce (Akbari, Moslehi, Fathi, & Bozorgmehr, 2007). Despite all the opportunities arising from the formation of the cyber space, unfortunately this space has many injuries and challenges (mentioned in previous chapters), in terms of the damage caused by those quantitative studies. In this chapter we will speak about how one can deal with these injuries and challenges.

In the last decade, the emergence of virtual and web technologies has had a significant impact on the educational and learning routines. Some online educational sites such as Wikipedia and MOOC and social and virtual media also have made a huge change, they have enabled students to understand the complex scientific and educational concepts in a simpler way and easily access these educational materials. Some people argue that cyber tools offer a kind of intrusive technology that can change the educational process and damage it. Other people have long opposed this idea, and there are numerous universities that use online educational resources and provide students with educational materials of MOOCs. For example, among educational websites that affect the educational process, one can refer to wikis that represent the transformation of the educational process and transform the learning environment from a teacher-centered environment to a learner-centered environment. Wikis that have a large amount of knowledge and information, have a significant partnership and participation in it, with high

© Springer International Publishing AG, part of Springer Nature 2019   93
S. H. Sadeghi, *Pathology of Learning in Cyber Space*, Studies in Systems,
Decision and Control 156, https://doi.org/10.1007/978-3-319-91449-7_6

educational capacity, have become today a powerful online education tool. Despite the fact that most educational sites promise to provide personalized learning experience, most students and teachers face a lot of problems when examine educational contents in cyberspace to find high quality content in this space. For example, in a recent study, students stated that they could not understand the complex formulas and some scientific methods because of lack of familiarity with computer science and mathematics, and now they believe that there are few online systems and limited websites that can help them in a good way (Liu, Kazmer, Twidale, Hara, & Subramaniam, 2015).

Learning Feedback in Cyberspace

Information and communication technology has created tremendous changes in all aspects of human life. Since the late 1990s, the technology has enjoyed extensive facilities, due to the lack of time and space constraints along with the perception of the role and place of human resources in the innovation of organizations, has led to the emergence of new methods in learning such as e-learning. The development trend has been accelerated over the last decade, with the development of information and communication technology over the past decade. In such a space, education and training are provided with a new model, and the subject of e-learning is given more attention. Internet technology and access to the cyber space of the human being enables anyone, anywhere, anytime, with a cheap method, anywhere in the world. Such an excellent communication tool has been able to transform education and move it beyond the place, time, and political and geographical boundaries (Wu & Wu, 2008). In this regard, with the passage of time and the formation of various developments in the field of various technologies, methods and forms of education have changed. This trend has been such that over time, new forms of learning have been provided. Over the past 150 years, computer communication has been the most fundamental change in the context of human communication. The increase in personal computers along with the Internet has caused rapid changes in the community. Electronic communications and digital networks are changing the way we work, and the transformation of our personal and entertainment relationships. This change in structure has had a tremendous effect on learning needs and opportunities (Anderson, Garrison, & Archer, 2004). Internet-based systems in which multimedia content (text, audio, video, video-related, and computer-related content) is provided electronically via computers, along with access to databases and e-libraries. In these systems, a teacher with a student, a student with a student, one-to-one or one to several people can interact simultaneously or non-simultaneously via e-mail, computer conferencing, electronic announcement boards, etc. (UNESCO, 2003). This has created the idea of access to education for all. Ideas that represent a fair environment for everyone, in which everyone, regardless of their spatial and social status, can access educational facilities. Today, cyberspace has made it possible for students and others to go to education through cyberspace. This provides a platform for everyone who have access to formal education, informal and hidden resources, and on the other hand the ability to provide educational objectives on a large scale. Of course, in terms of e-learning, it should be noted that as long as we have access to some information for sending e-mail or circulating in a cyberspace, this information affects our instinctive learning for specific issues. Although some scholars arguing that finding information is not a category of education and this view may be true, but if we value and understand the information we seek, we may have learned that; this is not training, but learning (Shih, Feng, & Tsai, 2008). Accordingly, advances in information technology and its combination with community change have created new ideas for learning. These dramatic changes have an enormous impact on learning and teaching systems. Beneficiaries in the learning and learning paradigm need a capable and supported learning environment through appropriate resource design. To stay in such a global competitive market, education providers need to develop efficient and effective learning systems to meet the needs of the community. Therefore, there is a huge demand for effective, easy-to-use, flexible, well-designed, learner-centric, distributed, and equipped educational environments. Internet technologies have fundamentally and rapidly transformed the economic context with dramatic changes (Dolly Goldenberg & Carroll Iwasiw, 2005).

When the communication becomes an important factor and effective in e-learning, the student and the teacher will have benefits together in a mutual communication. This communication will easier the learning, and will create opportunities for their learning expansion, harder connection between them, easier reaching in their goals and finally and overall positive experience.

In this regard, the formation of these developments and the creation of e-learning spaces have provided feedback in the cyberspace. These feedbacks are displayed by users in various formats. For example, creating a blog with educational goals in cyberspace, accompanying the individual in providing information about the social network created in the cyberspace, providing solutions for better formal education, developing and applying information and communication technology in educational systems, etc. It should also be noted that this type of learning is through the use of cyber space and it provides the basis for the formation of Internet addiction, the continuous presence of a person in the cyber space, and so on. Based on this, the formation of this vast educational environment, along with its high facilities and capacities, has led people to enter this environment with different goals and purposes and try to use their facilities to induce their goals and beliefs to others. They will create a kind of learning in them so that feedback from these cyber-learning can also enhance more and more their goals and objectives, and also in some way they can develop and extend the target in the cyberspace (Fig. 6.1).

Fig. 6.1 Feedback in cyberspace

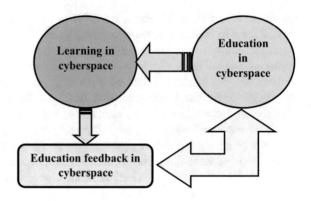

Learning Feedback Through Real-World Cyber Space

As already mentioned, the virtual world or cyberspace is another form of real world. For this reason, these two worlds are well comparable to each other. In fact, their primary elements are common, and only the nature, gender and role of these elements are different. The use of phrases such as virtual government, e-mail, e-learning and e-commerce, Internet commerce, virtual friendship, and so on, all of which primarily refer to cyber space, show the similarity between the elements of each of these two spaces. Although the development of communication technologies and the formation of cyber space has had many positive results and impacts on the advancement of communities in real space, it is impossible to ignore the negative effects of this space on real space. To respond to such a wide range of training with flexibility in time and space, content, process and training interactions, the use of new educational technologies, especially e-learning, are essential. E-learning and the use of knowledge distributed and facilitated by electronic technologies. In an e-learning environment, learning is more self-motivating and self-learning (Shih et al., 2008). Individuals or groups that are opposed to the political, social, or economic structure of the community are struggling to deal with the existing structure by publishing anti-cultural behaviors over the Internet. Hence, these groups or groups may, by questioning the current laws of the country, provide grounds for non-acceptance and abandonment of these laws among the people, especially the youth. Some other websites related to these individuals and groups try to promote ethnic and religious conflicts in the country and encourage them to violate the rule of the country. The charm of modern technology may make some people so enthusiastic that without knowing and evaluating enough, take practical steps that both hurt the training and make the work of the new technology useless and ineffective. The prevalence of new terms and words, new ceremonies, such as the Valentine's Day, are another influences of learning and learning in the cyberspace on real space.

One of the other benefits of e-learning, especially in its formal form, is to increase the cost of learning for individuals. Another common slogan that addresses the benefits of e-learning is to reduce training costs. While this slogan has been achieved in many universities in the world, it can be seen that in some countries, including our own country, the cost of e-learning courses for students is higher than other types of education. Of course, e-learning reduces costs for universities, because they can greatly increase their capacity by spending much less than traditional classes. However, reducing costs for students should also be important. Electronic learning although reduces the cost of travel for students, but on the contrary, it is a heavy sum. On the whole, in this regard, teaching and learning in cyberspace has affected various forms of real space. Apart from the damage caused by these trainings, as mentioned in the last chapter, this kind of education has created many opportunities for people to continue their education and transform their lives in real life.

How to Deal with Injury

Most studies on cyberspace describe the various opportunities and facilities of this space in various fields. In this regard, in order to be able to cope with the damage caused by training in its various forms in a cyberspace, and to minimize its effects, a comprehensive policy and plan should be adopted in this regard. Because it cannot be treated or damaged in a scattered and unplanned way. To deal with the damage and the challenges posed by this space, we must do the following:

Step one: Accurate and complete cyber space.

The first step in identifying the cyber space is to really believe in the existence of injuries. We must remove cotton from our ears, and if we are asleep, we wake up from sleep. And, according to the Supreme Leader, the revolution believed in the danger of cultural invasion, and we believe that the enemy has brought the scene of war to our homes in order to eliminate the religious and national identity of our people and collapse their siege. We need to call it a war with us and let everyone know about this by playing it. And provide the people with the necessary tools. One of the most important method to encounter this thread is the full knowledge of possible tools. The most important of these tools, which are used today for cultural invasion and other damages, is the cyber space. So we have to analyze this space well in order to know it completely. Identification and management are two very different things. The world may be complex, but if this recognition is to meet the needs of the community, then all the processes of the state should be simplified. Adaptations are essential, even if they are considered to be temporary political measures. With the full knowledge of cyber space, we have taken the first step in dealing with the damage caused by this space.

Step two: Train qualified and experienced staff to advance goals and understand the challenges.

After fully understanding the cyber space, we need to train forces for planning and shaping our goals in cyberspace and recognize the challenges and damages caused by cyberspace to enable us to build the necessary infrastructure in the cyberspace We create educational environments with the desired goals in this space and prepare the ground for the arrival of individuals and users.

Step three: Awareness of the people through the mobilization of all facilities.

Before entering the cyberspace, the capabilities and capacities of the cyber space on the one hand, and the limitations and challenges of this space on the other, should be identified to the public, in order to bring the people of the community with clear view of this space and the challenges and injuries. This may be created in this space through different educational resources for them. Among the features that you can use them to make people increase the level of awareness among people are of the following:

1. Sound: Fortunately, we have a good location for the people's attention to the radio and television networks, and we must use this opportunity and give the people the necessary knowledge about the cyber space
2. Use of journals and newspapers (through co-ordination and planning with the Ministry of Guidance)
3. Use of enormous education facilities
4. Using the facilities of universities and specialty fields to hold specialized conferences
5. The use of mosque and clergy capacity is very influential
6. Use of other cultural instruments such as cinema and theater
7. The use of educational facilities for military forces that all boys inevitably perform duty in the system are duty-bound
8. Assist in the capabilities of NGOs and NGOs …

Step four: Culture.

One of the other policies to be pursued in the field of education and culture should be comprehensive education, which today it means (in a society alone) education is not just a family and a society that is responsible for culture, but also many other institutions are part of it. Different ways are transmitting different concepts to their audience, which has a much broader range in cyberspace. This kind of transfer of culture through the cyberspace to people is carried out with different intentions and purposes, and on the other hand, this space is under the control of a group of leading countries that have the culture that is governing this space. To avoid such transfer (of culture), one society should, after informing people about the opportunities and challenges of the cyberspace, enhance the cyber cultural education resources with their own specific goals and culture and pave the way for the transfer of their cyberspace to users.

Step Five: Promoting the culture of proper use.

After we have created backgrounds and infrastructures for the transfer of the culture by strengthening the various educational resources in the cyberspace, we need to promote the proper use and utilization of resources in cyberspace among people and so on so that people use the right culture without encountering the challenge and damage of the capacities and capabilities of the cyber space.

Step Six: Develop the required laws and regulations.

Despite the fact that the use of cyber-space is rightly used, there is a risk of errors and other things in this space, and the causes of injury and challenges for themselves, others and communities may be inadvertently or in a planned manner. In order to prevent these kinds of injuries in this space, national laws and regulations should be imposed to punish both individuals and educational environments that cause injury and the lines

Red for other people in this space.

The other steps to deal with cyberbullying damage can be summarized as follows:

Step Seven: Managing Cyber Spaces.

Step Eight: Compile and deliver rich content in accordance with the desired goals.

Step Nine: Strengthen the ideological foundations of people in society, especially young people.

Step Ten: Strengthen national capacities in cyberspace (build games, website development, etc.)

Step Eleven: Create the background and ability to manage and use the family from the cyberspace.

Step Twelve: Formation of Cybercrime Police.

Of course, it should be noted that it may be possible to add other steps to the next steps or to move them. The purpose here is to express the ways in which they can reduce the damage caused by learning in cyberspace. Below is an example of how to reduce the cyber space damage to better understand the subject:

Enclosed garden

One of the strategies to prevent learning and injuries in the cyberspace is the enclosed garden. Here, the emphasis is on creating a safe area where learners and users can play without outside world involvement and benefit from training with the goals we seek. Within the fences, everything is safe and secure, and therefore there is no challenge in providing training for specific purposes. The enclosed garden covers only the right material. These materials often include curriculum resources, teacher software packages, and the equivalent of classroom Internet resources, content for reinforcing lessons and other resources with desirable goals. In this situation, the result of the work is completely assuring and any link and connection with the outside is banned and then the educational goals are realized and the field of transfer and learning is provided arbitrarily.

Capacities for Dealing with Injury

Social issues are not one-dimensional and always come from multiple social factors, and the solution to these issues should be sought in the correction of various factors, which is not easy, because the approach must be scientific. In a scientific approach to solving problems, firstly there must be the mode of view or "self" encounter with issues. In other words, modifications must start from "self", and this is not easy, but practical. Moving towards a science of living is a movement that will make two communities of the east and the west closer and blended. In this regard, in order to cope with the injuries and challenges posed by education and cyber space learning, before any action, we must identify and assess our capabilities in relation to that injury, and then we will identify ourselves in different ways to prevent and counteract that injury or challenge. In this regard, sociologists believe that instead of violent actions, there should be a plan for cultural promotion. On the basis of this, the following strategies can be used to build capacity to cope with the harm caused by training in cyberspace:

- Creating a sense of responsibility in people

The Berners lip team believes users are responsible for maintaining security and reducing cybercrime damage, "it's very important to know that the cyber space consists of content provided by users, to make these users who create the web" he says. Users are people who read, people who teach children how to use the web, who put information on the web, and especially those who create links on the web. The web does not force you to accept an opinion that is against your wishes. If you are worried that your child will seek low-level information, teach him or her about this. Teach him what kind of information to choose and how to value the information (Gray, 2004). Accordingly, in order to reduce the harm caused by learning in cyberspace, people have to create the feeling that they are more aware of the information they use within the cyberspace than the information they use in other spaces. And as a member of this space, efforts should be made to reduce these types of injuries and to isolate individuals and groups that provide different grounds for the challenge.

- Creating mutual interaction of the responsible institutions with the authorities of the websites and internal blogs

In this strategy, the government and authorities should create a space in which the interactions of the responsible institutions with the authorities of the domestic and human rights websites and blogs become more and more intimate. By doing so, they also strengthen the field of cyber space with content and training for their intended purposes, and they may on the other hand, knowingly damaging and challenging it, through the provision of harmful and challenging educational environments.

- Creating grounds for strengthening religious beliefs in individuals

Religion strengthens will in believers with its religious rules and puts the core of avoiding deviations into them, and leads to a high level of discipline, because discipline is a matter of commitment. The burden does not fit into discipline, and once discipline becomes overwhelming, one of the ways of penetrating deviation is closed. So deviant people are those who are outrageous about any kind of legitimate order. The health of the community depends on health of one person. If people are truly believers of the religion, they will have a clean, distant, and devious society.

The community of people who have the characteristics of religious beliefs leads to the formation of a healthy society in which individuals respect the laws, rights, and interrelationships of each other and regard justice as a sacred one. They trust and rely on each other. In this society, they do not hurt or injure others and they are responsible for themselves in different ways in front of others. In this regard, by strengthening religious beliefs in individuals and communities in different ways, by creating different educational resources in the cyberspace and real space, it is possible to avoid the challenges posed by the formation of educational environments with malicious purposes, as well as reduced exposure to cybercrime in these settings.

- Reinforcing and relying on national and local culture

The national and local cultures of any society have high abilities and capacities in dealing with various injuries and challenges. These cultures have been transmitted through generations and have grown in different ways, shaping the thoughts and beliefs of individuals about how to deal with others and society, and provide them with solid frameworks. Unfortunately, the formation of a breakdown between generations in societies, on the one hand, and the development of media, information and communication tools, especially the Internet, on the other hand, have caused a profound cultural change in lives of many people. The frameworks and the cultural backers of individuals that have been formed over many years and has faced a challenge that, in dealing with life's issues, lacks the essential capabilities that make them vulnerable to the problems of everyday life and its requirements. Accordingly, in order to prevent the spread of these kinds of injuries and challenges, we must eliminate the gap between generations and strengthen the national and local culture among people in the community to deal with these kinds of injuries.

- Creating e-literacy in society

The electronic literacy of human beings in society allows them to not passively pass through the events and transformations through cyberspace, and they are able to recognize this environment and have a clever confrontation with their injuries and challenges. This can be a source of self-esteem and a sense of personal commitment to the entry and use of the cyber-space, enabling them to recognize and cope with the challenges of the cyber-space. It also allows community members to prevent the harm of children and adolescents in this way to a large extent.

- Strengthening the foundations of families

As mentioned in the previous chapter, families have an important role in reducing the harm caused by education and learning in cyberspace. So, in a research on Internet communication in life, the role of perceived social support and loneliness in using the Internet has been done. It has been concluded that if the family does not play its role well as a source of perception of social support. A person will experience a feeling of affective family loneliness, and as a result, to fill his emotional vacuum and gain emotional support—which is not provided by the family—will shelter Internet environments and dangerous use of cyber space. Accordingly, in order to prevent the dependence of individuals, especially children, adolescents or other family members on cyberspace, and the reduction of their injuries, families should be able to minimize the emotional weakness among their members and strengthen the family's foundations. Of course, governments can play a significant role in helping to strengthen the foundations of the family in the present era. Dr. Jay Katz, a professor at the University of Toronto, says that teenagers who regularly use the Internet are usually isolated and intimidated, and to find someone who can give them consultation about threats and abuse, they get into trouble. To overcome this problem, teachers, school authorities and parents should have sufficient knowledge about the prevalence of cyber-related injuries to inform students and cyberspace users of how to use cyber space safely and appropriately. Additionally, since adults are generally less knowledgeable about the use of the Internet than students are, they must continually update their knowledge (Kornblum, 2007).

Overview

Recent decades, and the subsequent transformation of human communication, have had a tremendous impact on the needs and opportunities of human learning. This is due to the fact that the existence of various and varied content in the Internet space in the form of text, audio, image … along with access to databases and libraries and e-learning classes, and so on, i.e., a vast environment available to all different spectrums of society has put education into learning. In another word, this means creating equal opportunities for all to raise the quantitative and qualitative levels of individuality as well as social knowledge and skills.

Of course, it should be noted that along with the development of these trainings at the community level and the existence of positive effects from it, feedback and damage (both in the cyberspace and in the real space of society) have emerged.

This chapter is devoted to step-by-step strategies for dealing with these injuries.

Chapter 7
Conclusions

Introduction

Reviewing the history suggests that technology has always come to the aid of mankind and helped him improve work and increase efficiency. In this regard, the category of teaching and learning has not been overlooked and has been stepped up with the advancement and development of day-to-day technologies. It has transformed itself into using technology and even has brought new forms. In ancient times, the learning network was only limited within the family of birth and life, and with the expansion and formation of tribes and communities, the education and training network was developed relatively to the past. Nevertheless, the scope of education and learning was not widespread. With the evolution of human societies and the growing human needs and consequently the development of various technologies, gradually the idea of establishing educational institutions for solving individual and social skill needs and the transfer of experiences and knowledge as a pressing need among people were gradually developed. At the same time, with the development of other educational sectors in society and the development of formal institutions, each of which has a part in teaching and learning, the family still retains some of the basic educational functions, and the social relations network also has its own teaching functions. Found in this process, the field of education and learning has become a vast area that has created various educational institutions that, in their own way, generally or in the field specialized areas have undertaken training and learning activities. The new needs, the breadth of demand, the development of tools, and in particular the advent of new technologies, have transformed the supply of education into enormous changes. But none of these developments, like the developments of recent decades in the development of ICT, have failed to transform education and learning. Developing information and communication technology and the formation of a cyber space have paved the way for a major transformation in education and learning in communities. The cyberspace, with its high capacities and capabilities, was able to transform, in different ways, all the arenas and

© Springer International Publishing AG, part of Springer Nature 2019
S. H. Sadeghi, *Pathology of Learning in Cyber Space*, Studies in Systems,
Decision and Control 156, https://doi.org/10.1007/978-3-319-91449-7_7

resources of learning and education in advance, in the face of past developments that created a new source alongside past resources of education and learning. This provided the ground for even their destruction. For example, the cyber space with its high abilities and capacity has transformed and eroded the foundations of the first and most important educational entity of the people, and has transformed various forms of education and learning from this educational institution. This has made the present century, with all its economic and social advancements which have not had the same fruit for all, to become one of the important ages during the life of mankind. In this regard, it is necessary for all people to pay attention to the goals and means of teaching and learning, and given the increasing trend of developing capacities and facilities of cyber space as a transformation that has changed other educational institutions. They need to know the opportunities for learning and education that are vital to human beings. However, in the process of developing cyberspace and expanding its capabilities and functions, researchers and experts have tried to recognize the opportunities of learning from the cyberspace, but unfortunately, in this process attention to the injuries and challenges caused by training and learning in the cyber space has been less widely considered. In this regard, the present book attempts to cover as much as possible the damage caused by the growing trend of education and learning through cyber space in various sectors of government damages, community damage and … This process is presented in the following general form (Fig. 7.1).

As stated above, human beings have always tried to transfer their knowledge and experiences to subsequent generations. This human need and endeavor to immortalize with stone writings and then writings on the animal's skin until the invention of the paper continues to this day. Of course, its most important part is formal education with the participation of the teacher and the formal education and training institutions. This kind of transfer of experience and knowledge has been accomplished for many years. But today, with the formation of the cyber space, fundamental changes have taken place. The cyberspace, with its high capacities and capabilities, has been able to play a major role in teaching and learning.

The emergence of new technologies to support cyber-space applications has led to the emergence of new types of vulnerabilities, especially in the education and learning sector. In this regard, the widespread presence of community members, especially students in the cyberspace, is not wise without regard to what we lose and what we get. This space has many capacities and capabilities that shape learning experiences in a different way and their perspective on the world. So, in teaching and learning, it has become commonplace for individuals to participate in learning environments, which is informal education in these environments or hidden, would have acquired the knowledge and experience of others. But the cyber space has been able to create this environment in a variety of ways, regardless of the time frame and spatial framework, and put people in different forms under formal, informal or hidden training, or all three at the same time. This environment has been able to interact with the use of text, sound, photo, map, video, etc., and in shaping them in sunny environments, and in many cases expressing the educational goals in question, an environment of high interest in education. It also provides people with

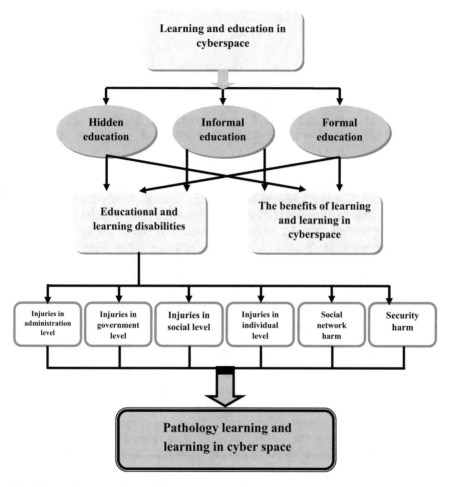

Fig. 7.1 Learning pathology and learning in cyber Space

learning and transferring their goals and knowledge to cyber-space users. The existence of different teaching and learning environments with different goals in cyberspace, which is easily accessible to individuals and users are easily placed in the control of education, has caused various damage to individuals and students.

These injuries enter users through various official sources such as academic or virtual schools, private centers, informal resources such as websites, blogs, ... or hidden resources of games, pictures and .... These injuries can be categorized in general in the following sections:

- Injuries at the government level
- Government-level injuries
- Injuries at the community level

- Injuries at the individual level
- Social network injuries
- Security damage

   These injuries, which result from education and learning in the cyberspace, affect the individual, society, government and government in various ways. Such harm, if neglected by the community and the government, will be very large and sometimes irreversible, since the psychological harm caused by the training in this space and the wrong use of the law would disrupt the behavior of the citizens and society. Finally, it avoids the desired educational goals and the development of human and religious personality. As a result, some harms occur such as the weakening and challenge of family foundation, depriving students of the benefits of ethical and educational relationships between teachers and students, students and faculty in cyberspace, avoiding communication, lack of development of social skills, induction of inappropriate behavior towards individuals. Invasion and cultural transfer through education in cyberspace create a challenge for governments and … including cybercrime injuries. These disorders make the citizens worse and disorientate their daily activities. The social and cultural damages that result from it make the members of the community alien to individual behavior with the family and to social behavior with other citizens and to a shaky and distorted state of affairs. The norms and values of the transcendental community of the decayed society have a sense of security and peace of mind. Meanwhile, its political damages undermine the authority and the sovereignty of the state, poses a serious challenge in creating national unity, social security and information security. Accordingly, in this book, we tried to identify and reconsider all the damages caused by cyber-learning education and learning. Despite the hardship that has been posed in this book on cyber training and learning, we should note that these new technologies, including the Internet and the formation of cyber space, are an inseparable part of life. It is now in front of us and has indeed many benefits and uses for us too. Now the use of cyber space in developed countries is part of the everyday life of the people of these countries and has been able to provide these countries (apart from the damages caused by these capabilities and high capacities, and given the interconnectedness and interdependence). At present, different nations are opposed to each other. Lacking this trend and creating a digital divide between countries with domination and influence in the cyberspace can in the near future have more serious disruptions and harms for national interests at the different levels especially in terms of education and learning. Among them, the status and role of educational institutions at different levels and forms from one person to another, from government to non-public, from formal to informal, in order to teach how to use this space and to recognize its damages to provide them with some knowledge to avoid it. The benefits of the country and the reduction of injuries and challenges posed by cyberspace, especially in the education and learning sector, are very important. Based on this, we will identify and strengthen existing capacities for optimum use of this space and build capacity in areas where we are weak in order to make better use and reduce vulnerability. By this we can prepare it using our authority according to the development of cyberspace in different areas. Authority in this newly formed space means an active participation in the production and distribution of information defined in the framework of predetermined strategies. In most cases, cyberspace users have the right to choose not passively. But if we are in the consumer information and backward group, we take a passive position because of the fact that it is not in the process of its production and distribution. We can expand knowledge, culture and so on, where cyber has a role. In general, political, cultural, scientific and economic statures on the cyberspace means the interference with the masses, the re-ordering of the values, behaviors and identities of the nations through which cyber-space damage can be avoided as much as possible. On the other hand, he oversaw this area as a superior power.

In the end, we have to clarify the importance of recognizing, coordinating and changing ourselves with the changes made by ICT in general and the formation of a cyber-space in a specific way, according to Charles Darwin's famous statement on survival, let's face it:

"These are not the strongest and most intelligent species that remain, but those who are most likely to respond to changes".

References

Adams, J. (2017). *The next world war: The warriors and weapons of the new battlefields in cyberspace.* Retrieved from https://dl.acm.org/citation.cfm?id=3165144.

Agarwal, P. (2009). *Indian higher education: Envisioning the future.* New Delhi: Sage Publications India.

Akbari, M., Moslehi, A., Fathi, S., & Bozorgmehr, S. (2007). *Identification and classification of national and international experiences in the development of electronic commerce.* Mumbai: Publications of the Institute for Business Studies and Research.

Alraimi, K. M., Zo, H., & Ciganek, A. P. (2015). Understanding the MOOCs continuance: The role of openness and reputation. *Computers & Education, 80,* 28–38.

Anderson, T. (2004). Towards a theory of online learning. In *Theory and practice of online learning* (Vol. 2, pp. 109–119). Athabasca: AU press.

Anderson, T., Garrison, D. R., & Archer, W. (2004). Critical thinking, cognitive presence, computer conferencing in distance learning. *Tersedia.* http://www.communityofinquiry.com/files/CogPres_Final.pdf.

Aparicio, M., Bacao, F., & Oliveira, T. (2014). *Trends in the e-learning ecosystem: A bibliometric study.* Lisbon: Elsevier.

Bagchi, A., & Bandyopadhyay, T. (2018). *Role of intelligence inputs in defending against cyber warfare and cyber terrorism.* Retrieved from http://www.gtcenter.org/Archive/2016/Conf/Bagchi2341.pdf. Last accessed May 14, 2018.

Barnes, S. J., & Pressey, A. D. (2014). Caught in the web? Addictive behavior in cyberspace and the role of goal-orientation. *Technological Forecasting and Social Change, 86,* 93–109.

Borges, J., Justino, E., Gonçalves, P., Barroso, J., & Reis, A. (2017, April). Scholarship management at the university of Trás-os-Montes and Alto Douro: An update to the current ecosystem. In *World Conference on Information Systems and Technologies* (pp. 790–796). Cham: Springer.

Bouchrika, I., Harrati, N., Mahfouf, Z., & Gasmallah, N. (2018). Evaluating the acceptance of e-learning systems via subjective and objective data analysis. In *Software data engineering for network elearning environments* (pp. 199–219). Cham: Springer.

Broadhurst, R. (2017). Cybercrime in Australia. In *The palgrave handbook of Australian and New Zealand criminology, crime and justice* (pp. 221–235). Cham: Palgrave Macmillan.

Brown, J. S., & Duguid, P. (2000). Mysteries of the region: Knowledge dynamics in Silicon Valley. *The silicon valley edge* (pp. 16–45). Stanford, CA: Stanford University Press.

Callan, V. J. (2009). *How organisations are using e-learning to support national training initiatives.* Queenland: Department of Education, Employment and Workplace Relations.

Chakraborty, S. (2017). *Module functioning of computer worm, PC virus and anti virus programs*. West Bengal: MAKAUT.

Chauhan, A. (2014). Massive Open Online Courses (MOOCS): Emerging trends in assessment and accreditation. *Digital Education Review, 25*, 7–17.

Chin, A., & Jacobsson, T. (2016). TheGoals.org: Mobile global education on the sustainable development goals. *Journal of Cleaner Production, 123*, 227–229.

Connolly, T. M., Stansfield, M., & McLellan, E. (2006). Using an online games-based learning approach to teach database design concepts. *Electronic Journal of E-learning, 4*(1), 103–110.

Dean, L. A. (Ed.). (2006). *CAS professional standards for higher education*. Council for the Advancement of Standards in Higher Education.

Di Angelis, G. (2004). *Cybercrime*. Philadelphia. Pa: Chelsea House.

Dolly Goldenberg, R. N., & Carroll Iwasiw, R. N. (2005). The effect of classroom simulation on nursing students' self-efficacy related to health teaching. *Journal of Nursing Education, 44*(7), 310.

Dron, J., & Anderson, T. (2014). *Teaching crowds: Learning and social media*. Athabasca: Athabasca University Press.

Egenfeldt-Nielsen, S. (2005). Can education and psychology join forces. *Nordicom Review, 26*(2), 103–107.

El-Hmoudova, D. (2015). Motivation and communication in the cyber learning environment. *Procedia-Social and Behavioral Sciences, 191*, 1618–1622.

Fadel, R., Deniau, T., Fullerton, G. B., Inoue, A., Stephens, T., & Carlhian, A. (2017). *U.S. Patent No. 9,563,488*. Washington, DC: U.S. Patent and Trademark Office.

Ghorbani, A., & Ghorbani, A. (2014). Investigating computer crimes in cyberspace. *Kuwait Chapter of the Arabian Journal of Business and Management Review, 3*(10), 399.

Giddens, A. (2009). *Sociology* (1st ed). Cambridge: Polity Press.

Gordon, S., & Ford, R. (2002). Cyberterrorism? *Computers & Security, 21*(7), 636–647.

Graham, C., Thompson, C., Wolcott, M., Pollack, J., & Tran, M. (2015). A guide to social media emergency management analytics: Understanding its place through Typhoon Haiyan tweets. *Statistical Journal of the IAOS, 31*(2), 227–236.

Gray, B. (2004). Informal learning in an online community of practice. *Journal of Distance Education, 19*(1), 20.

Hallikainen, H., & Laukkanen, T. (2018). National culture and consumer trust in e-commerce. *International Journal of Information Management, 38*(1), 97–106.

Hills, R. M. (2005). Corruption and federalism: (When) do federal criminal prosecutions improve non-federal democracy? *Theoretical Inquiries in Law, 6*(1), 113–154.

Hong, J. C., Tai, K. H., Hwang, M. Y., Kuo, Y. C., & Chen, J. S. (2017). Internet cognitive failure relevant to users' satisfaction with content and interface design to reflect continuance intention to use a government e-learning system. *Computers in Human Behavior, 66*, 353–362.

Houston, J. E. (2001). *Thesaurus of ERIC descriptors*. Westport: Greenwood Publishing Group.

Jacobson, M. J., Kim, Y., Lee, J., Kim, H., & Kwon, S. (2005). Learning sciences principles for advanced e-learning systems: Implications for computer-assisted language learning. *Multimedia Assisted Language Learning, 8*(1), 76–115.

Jacobson, M. J., & Spiro, R. J. (1994). A framework for the contextual analysis of technology-based learning environments. *Journal of Computing in Higher Education, 5*(2), 3–32.

Kenney, M. (2015). Cyber-terrorism in a post-stuxnet world. *Orbis, 59*(1), 111–128.

Kim, C., & Keller, J. M. (2008). Effects of motivational and volitional email messages (MVEM) with personal messages on undergraduate students' motivation, study habits and achievement. *British Journal of Educational Technology, 39*(1), 36–51.

Kornblum, J. (2007). The net is a circuit of safety concerns. *USA Today*.

Korzinov, V., & Savin, I. (2017). General purpose technologies as an emergent property. *Technological Forecasting and Social Change*. Karlsruhe: Elsevier.

Lee, E. A. (2006). Cyber-physical systems-are computing foundations adequate. In *Position Paper for NSF Workshop On Cyber-Physical Systems: Research Motivation Techniques and Roadmap* (Vol. 2, pp. 1–9).

Li, M. (2017). *The wars in your machine: New developments in trojan virus engineering.*

Liu, X., Kazmer, M., Twidale, M., Hara, N., & Subramaniam, M. M. (2015). Education in the cyberlearning era: New challenges, opportunities, and applications. *Proceedings of the Association for Information Science and Technology, 52*(1), 1–3.

Luckmann, T., & Berger, P. (1964). Social mobility and personal identity. *European Journal of Sociology/Archives Européennes de Sociologie, 5*(2), 331–344.

Lumby, C., Anderson, N., & Hugman, S. (2014). Apres Le Deluge: Social media in learning and teaching. *Journal of International Communication, 20*(2), 119–132.

Mack, E. (2017). *Spy satellite SpaceX launched might buzz the space station.* [online] CNET. Available at. https://www.cnet.com/news/spacex-spy-satellite-usa-276-nrol-76-international-space-station-nasa/. Accessed Mar 1, 2018.

Manca, S., & Ranieri, M. (2016). Facebook and the others. Potentials and obstacles of social media for teaching in higher education. *Computers & Education, 95,* 216–230.

Mayadas, F., Miller, G., & Sener, J. (2015). *Definitions of elearning courses and programs: Version 2.0.* NewYork: Online Learning Consortium.

Mkrttchian, V., Gevorgian, S., Shoukourian, S., Gasparyan, F., Vardanyan, R., Poghossian, A., et al. (2018). Student competence. *Handbook of Research on Students' Research Competence in Modern Educational Contexts, 1.*

Owen, T., Noble, W., & Speed, F. C. (2017). Virtual violence: Cyberspace, misogyny and online abuse. In *New perspectives on cybercrime* (pp. 141–158). Cham: Palgrave Macmillan.

Özdemir, E., & Aydın, S. (2015). The effects of blogging on EFL writing achievement. *Procedia-Social and Behavioral Sciences, 199,* 372–380.

Polański, P. P. (2017). Cyberspace: A new branch of international customary law? *Computer Law & Security Review, 33*(3), 371–381.

Pourghaznein, T., Sabeghi, H., & Shariatinejad, K. (2015). Effects of e-learning, lectures, and role playing on nursing students' knowledge acquisition, retention and satisfaction. *Medical Journal of the Islamic Republic of Iran, 29,* 162.

Prensky, M. (2006). Learning in the digital age. *Educational Leadership, 63*(4), 8–13.

Robinson, K. (2010). Changing education paradigms. *RSA Animate, The Royal Society of Arts, London.* http://www.youtube.com/watch.

Rolando, L. G. R., Salvador, D. F., & Luz, M. R. (2013). The use of internet tools for teaching and learning by in-service biology teachers: A survey in Brazil. *Teaching and Teacher Education, 34,* 46–55.

Rosenberg, M. J. (2005). *The future of e-learning. ASTD's learning circuits webzine.* NewJersey: McGraw-Hill.

Sadeghi, S. H. (2018). *E-learning practice in higher education: A mixed-method comparative analysis.* Sydney, NSW: Springer.

Saha, C. N., & Bhattacharya, S. (2011). Intellectual property rights: An overview and implications in pharmaceutical industry. *Journal of Advanced Pharmaceutical Technology & Research, 2* (2), 88.

Shih, M., Feng, J., & Tsai, C. C. (2008). Research and trends in the field of e-learning from 2001 to 2005: A content analysis of cognitive studies in selected journals. *Computers & Education, 51*(2), 955–967.

Shopova, T. (2011, June). E-learning in higher educational environment. In *Proceedings of International Conference The Future of Education* (pp. 16–17).

Singer, P. W., & Friedman, A. (2014). *Cybersecurity and cyberwar: What everyone needs to know.* NewYork: Oxford University Press.

Squire, K., & Jenkins, H. (2003). Harnessing the power of games in education. *Insight, 3*(1), 5–33.

Stensaker, B., Maassen, P., Borgan, M., Oftebro, M., & Karseth, B. (2007). Use, updating and integration of ICT in higher education: Linking purpose, people and pedagogy. *Higher Education, 54*(3), 417–433.

Stockley, D. (2003). *How OE-learning definition and explanation (E-learning, online training, online learning) online learning is revolutionizing K-12 education and benefiting students.* Retrieved 21 January 2015.

Stoltz, T., Weger, U., & da Veiga, M. (2017). Higher education as self-transformation. *Psychology, 7*(2), 104–111.

Strauss, E. (2002). Hepatosplenic schistosomiasis: A model for the study of portal hypertension. *Annals of Hepatology, 1*(1), 6–11.

Suarez-Tangil, G., Dash, S. K., Ahmadi, M., Kinder, J., Giacinto, G., & Cavallaro, L. (2017, March). DroidSieve: Fast and accurate classification of obfuscated android malware. In *Proceedings of the Seventh ACM on Conference on Data and Application Security and Privacy* (pp. 309–320). New York: ACM.

UNESCO Advisers. (1990). *Educational planning process.* Paris: United Nations Educational, Scientific and Cultural Organisation.

UNESCO Advisory Group. (2003). *Educational planning process.* Paris: Print thirteenth. United Nations Educational, Scientific and Cultural Organisation.

United Nations Development Programme. (2014). *Human development report 2014: Sustaining human progress: Reducing vulnerabilities and building resilience.* Retrieved from http://hdr.undp.org/sites/default/files/hdr14-report-en-1.pdf.

Vali, I. (2013). The role of education in the knowledge-based society. *Procedia-Social and Behavioral Sciences, 76,* 388–392.

Wu, S., & Wu, L. (2008). The impact of higher education on entrepreneurial intentions of university students in China. *Journal of Small Business and Enterprise Development, 15*(4), 752–774.

Xu, Y., Yang, J., & Zhong, S. (2017). An online supervised learning method based on gradient descent for spiking neurons. *Neural Networks, 93,* 7–20.

Yeboah, J., & Ewur, G. D. (2014). The impact of WhatsApp messenger usage on students performance in tertiary institutions in Ghana. *Journal of Education and Practice, 5*(6), 157–164. Takoradi: International knowledge sharing platform (IISTE).

Zhang, D., & Nunamaker, J. F. (2003). Powering e-learning in the new millennium: An overview of e-learning and enabling technology. *Information Systems Frontiers, 5*(2), 207–218.

Zhu, Q. (2016). *Recent security issues in personal communication system.* Evanston: IEEE.

Printed in the United States
By Bookmasters